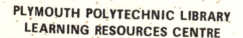

Planning of Water Quality Systems

Planning of Water Quality Systems

William Whipple, Jr.
Rutgers University

Lexington Books
D.C. Heath and Company
Lexington, Massachusetts
Toronto

Library of Congress Cataloging in Publication Data

Whipple, William, 1909-
   Planning of water quality systems.

   Bibliography: p.
   Includes index.
   1. Water quality management. I. Title.
TD365.W45   363.6'1   76-47336
ISBN 0-669-01144-4

Published simultaneously in Canada.

Printed in the United States of America.

International Standard Book Number: 0-669-01144-4

Library of Congress Catalog Card Number: 76-47336

Contents

List of Figures

List of Tables

Preface

This book was written because after a good many years spent in water resources work, of which the last ten have been devoted to water pollution control research, I have concluded that our national water pollution control effort is misdirected in that it is proposing to expend vast resources without commensurate environmental returns. This view is also held by most of my associates in various parts of the country who have intimate knowledge of current activities.

The book has no conscious ideological or political basis, but it is based upon three strongly-held beliefs:

1. That the preservation of an environment pleasing to man and useful to natural ecosystems remains an important national objective
2. That since the economic resources available to achieve our national objectives are limited in terms of manpower, materials and energy, the planning of water pollution control activities so as to minimize use of these resources is essential
3. That the United States deserves reasonably efficient administration in the interests of all the people, and that the processes of government should be re-examined from time to time

Business and political activists who would limitlessly exploit our land for economic development will find little to comfort them in these chapters, but no more comfort will be found by those environmental activists who see themselves as the protagonists of a holy war, to be carried on regardless of the cost to other societal objectives. The book is written to show how the conflicting environmental and economic interests can best be reconciled. This book also provides information on new scientific developments, applies social and economic criteria to policy review, indicates how present planning approaches should be improved, and points out waste, confusion and obfuscation.

The book is not written mainly for specialists, but for planners, administrators, policy analysts, and public interest groups, to give them a better understanding of complex scientific processes and the extent to which certain technologies are reliable. It does not attempt to cover all parts of a vast and complex subject with enough scientific detail to satisfy specialists in the various fields concerned, however, it may benefit specialists by informing them of fields other than their own. The book is based upon long experience in planning, which includes recent experience in organizing and participating in three areawide wastewater management plans. It is also based upon the experience, data and opinions of many associates in academic research circles, consulting firms, and various levels of government.

Some parts of the book are quite critical of our present national pollution control legislation. Hopefully, Congress will soon improve that legislation. However, our nation is so vast and so diverse that laws can only establish a framework within which specific plans can be developed, each suited to the region and the situations involved. The principles and criteria outlined in these chapters will be applicable to the development of such specific plans, subject to the constraints imposed by law.

Much of the technical knowledge and data cited comes from the research and knowledge of associates at Rutgers University, particularly Drs. Joseph V. Hunter, Samuel D. Faust, Harold H. Haskin, and Shaw L. Yu. Dr. Hunter and Dr. Bernard B. Berger helped greatly with technical review. The Office of Water Research and Technology, Department of the Interior, provided research support that helped in writing several chapters. Anna Klein did nontechnical editing, Lois Johnson prepared the drawings, and three loyal secretaries did the typing: Marjorie E. Krespach, Jean B. Potter and Muriel R. Katz. Many others whose ideas were used are cited in the references. For any inadequacies, errors and undiplomatic expressions, the author is solely responsible.

August 1976
Princeton, New Jersey

Planning of Water Quality Systems

1 Water Pollution, A National Problem Area

For every problem there is a solution—simple, neat and wrong.

H.L. Mencken

Clean rivers! The words conjure up pleasant associations of trees, canoes and sailboats, fishing, picnics and family outings, and a thousand other pleasant details. *Pollution* is another emotion-packed word, evoking thoughts of disagreeable sights, bad odors, the instinctive revulsion related to excrement, and perhaps the sickening memory of dead fish floating on foul waters, or ducks dying in oil spills. The phrase *pollution control* commonly comes across as a sort of imperative, like exterminating roaches in a kitchen or curing infectious diseases.

These "gut reactions" are healthy and sound and a credit to the decency and forthrightness of the American public. They provided important support for the great environmental movement of the late sixties and early seventies which launched the United States upon a pollution control program of truly enormous dimensions, aimed at completely eliminating the great evil of pollution, nationwide, by 1985.

Pollution control, however, is also a science, or rather an immensely complex technology which involves various fields of science and engineering and which has economic, social and cultural aspects that are not well understood. The vast program of pollution control undeniably started off in the right general direction, however, it seems to be moving haltingly, in some cases ineffectively, and in many cases with unnecessarily high costs. Decision makers must beware of being baffled and misled by the scientific jargon of pretentious manipulators, but also, they must not plunge blindly ahead in disregard of the plain warnings of the best scientific minds. It is worth an effort to understand what really is going on in this important field.

Background

The water resource development of the United States started out with separately planned and funded single-purpose programs for navigation and for irrigation in the west, and later, flood control and hydroelectric power. However, even before this time, the idea spread that single-purpose development of water resources was not in the public interest. A policy was established that planning was to be

1

comprehensive in that it should encompass all appropriate potentialities of the rivers in question, and also that projects should be multiple-purpose when such development would best fit the needs of the region and of the nation. In the 1950s and 1960s a great deal of public attention was paid to perfecting principles of comprehensive water resources planning and to establishing governmental mechanisms for carrying them out, culminating in the organization of the Water Resources Council.

Initially, the federal government was not concerned with water pollution control, which was considered a matter of state and local interest. In 1964, after some preliminary phases, a national system of water quality standards was adopted by the states, under strong federal influence. This act was meant to fill the major gap in the arrangements for comprehensive water resource planning under the Water Resources Council. However, this intended integration of federal planning did not develop productively. Instead, the Pollution Control Act of 1972 (PL 92-500) established decision making processes in water pollution control that are conducted independently of the comprehensive planning under the Water Resources Control. (Although called amendments, this is practically a new law. It will be referred to frequently in the text as the 1972 Act. The administrator of the Environmental Protection Agency did not become a full member of the council until October 1975.) The Water Resources Council and its constituent agencies have continued to function with the same responsibilities and procedures as previously, but they do not, in fact, plan effective pollution control. Comprehensive planning without inclusion of water quality control is severely handicapped. The "Principles and Standards" of the Water Resources Council now apply to less than 50 percent of water resources programs, including water quality control programs, when measured in terms of federal spending.[1]

What happened was that the environmental movement, which had developed during the late sixties, acquired sufficient momentum to initiate a vast new national program of water pollution control. The new program was politically based upon the environmental movement and its proponents separated it from all of the traditional approaches to planning of water resource development, in order that it might appear fresh and new to the public. This approach was highly successful in implementing what undoubtedly was a national consensus that a major move be made to clean up our river systems. At the same time, other established river development programs were suspended pending an in depth examination of environmental aspects under the National Environmental Protection Act (NEPA). Decision making under these conditions poses new problems and most analysts feel that some basic changes are highly desirable. In any case, it is necessary that principles of planning be applied, rather than relying upon the decision making approaches used so far.

The Current Water Pollution Control Program

The already considerable federal pollution program under the older legislation has been greatly accelerated under the 1972 Act. National water quality control expenditures were estimated by the Bureau of Census at $4 billion in FY 1973, of which $1.4 billion was federal.[2]

Net obligations, in millions of dollars, under the federal construction grant program were as follows:

FY 1972	$ 860 million
1973	2989
1974	2633
1975	4132

Most of the grants are made for public waste treatment plants, although sometimes construction of interceptor or outfall sewer lines can be covered. No grants are made for operations and maintenance expenses, industrial treatment facilities, or control of urban runoff.

The annual report of the Environmental Protection Agency for 1975 indicates that progress is being made nationally in improving levels of dissolved oxygen and reducing coliform counts in our waterways, but that improvement as regards toxics is scattered, and there is actually an increase of undesirable eutrophication conditions.[3]

Important aspects of our current pollution control strategy include the immense delegated powers of the EPA and the dependence for guidance and certain key decisions upon litigation. In contrast to previous water resource development programs which had to submit formal plans in each particular case and then rely upon congressional decisions and funding actions, the 1972 Act assigns general funding powers to the administrator, subject mainly to court constraints. The Act is worded largely in terms that are not definable either on any scientific basis or through practice in any profession. Waste treatment goals, as defined by the administrator, are described in terms of "best practicable" or "best available" technology, and are meant to be applicable throughout the entire United States. It is inevitable that such arbitrary determinations will be tested in the courts. The EPA is currently involved in hundreds of lawsuits, some of them involving very fundamental questions about the law's intent, particularly concerning effluent guidelines and limitations.[4]

Criticism of the 1972 Act from nongovernmental sources is widespread. Many state officials are highly critical, although their public remarks are generally muted for fear of offending an agency that disburses billions of dollars annually. The Water Pollution Control Federation (WPCF) supports the objectives of the act, but it concludes that "The Act and its application have not, in the main, begun to approach public expectation."[5] Suggested changes would give more decision making power to the states, particularly allowing them to correct "the failure of the Act and its current administration to consider adequately local conditions in imposing effluent limitations on discharges."[6] WPCF says that standards imposed nationally which minimize the importance of local conditions cause waste of limited resources. The claim is made that large cities such as Boston and Chicago have had to postpone action on critical combined sewer problems in order to comply with a secondary treatment standard which offers little hope of water quality improvement. Water management officials in Pennsylvania report that office time spent by technical and professional staff is increasing at six times the rate of field work.[7] The president of WPCF in 1974 was even more outspoken, concluding that the 1972 Act is not suitable for implementation in its present form. He said that everyone in the field has felt frustrations "as efforts and hopes are swallowed up in a quagmire of federally created red tape" in addition to state and local red tape.[8]

Regarding the EPA research program, the WPCF research committee[9] concluded that the nation and the objectives of PL 92-500 are being poorly served by present water pollution control research efforts: "Today's treatment technologies on which the private and public sectors are spending billions of dollars annually are expensive and often incapable of effecting levels of pollutant removal that protect water quality."[10] The committee recommended developing the following four aspects in order to reconcile environmental and economic demands:

1. Analytical tools for measuring and assessing the problem
2. Improved and more cost-effective treatment technologies
3. Environmentally more acceptable methods of disposing of pollutants removed from our waters
4. Management policies that assure optimum and equitable implementation of control strategies

The General Accounting Office also reported that research efforts in the water pollution fields are inadequate, and that the low current funding for research is the reason for the failure to develop technology for achieving established water quality goals.[11] Among deficient research areas cited are programs to determine how pollutants get into water, their effect, and what happens to them.

A consultant's study for the National Utility Contractors Association[12] found that the municipal sewage treatment grants program has "no goals, no deadlines, and no responsibility at the operating level, and is burdened by regulations which, when stacked together, reach as high as ten inches. The regulations take two weeks to read." It was found that in most cases there is a two- to four-year lapse between conception of a project and start-up of construction, with the total time before a facility can begin operation averaging six years. More than 60 percent of this time is used for the review procedures necessary to process the grant application. "As many as fifty-five people are typically involved in the development and review of project design and specification before a final obligation decision can be made." Locally, municipalities "are not responding quickly to the opportunity to use construction grants, even at 75 percent federal funding." When the money is obligated, the municipalities are often slow to let contracts and begin construction.

Some interesting points were raised by Dr. Joseph Ling on behalf of the Minnesota Mining and Manufacturing Company in testimony before the U.S. House of Representatives Committee on Public Works.[13] He presented data indicating that to remove 4,000 tons of pollutants from one of 3M's plants, more than 40,000 tons of natural resources would be consumed and approximately 19,000 tons of residues would be released to the air and the land. The plant currently provides secondary treatment and produces an effluent that met (as of August 1972) both state and federal standards. He considered that three important questions should be considered before requiring additional treatment in this situation: (1) Can the natural receiving waters safely assimilate the 4,000 pounds of water contaminants? (2) which is more detrimental to the total environment—4,000 tons of water contaminants or 19,000 tons of air and land contaminants? (3) is the expenditure of 40,000 tons of natural resources worth the net improvement to be gained in the natural receiving waters?

In this particular case, the pollutant removed (asbestos fibers) is considered potentially hazardous to human health, in fact, possibly carcinogenic, so that in this case its removal may well have been worth much more than the costs to society cited by Dr. Ling. However, there are no similar grounds cited for the general removal of pollutants required by PL 92-500. The general removal of very small quantities of polluting substances in refinements of advanced waste treatment may often require inputs of energy, manpower and scarce resources that far outweigh any environmental advantage obtained. On the other hand, there are obviously cases in which the removal or prevention of even very small concentrations of polluting substances may be necessary. The most telling criticism of the 1972 Act is that it makes insufficient provisions for distinguishing between the two cases.

State and Federal Water Quality Controls

Under federal legislation prior to 1972, all of the states classified streams, stipulating minimum water quality standards, in accordance with their planned uses. There are usually special requirements for trout streams (requiring low temperature, little turbidity, and high DO), and contact sports (especially low coliform counts), etc. At the time, these water quality standards were meant to constitute the direct and immediate goals of water quality planning and federal support programs. These standards are still in existence, though greatly de-emphasized by certain provisions of the 1972 Act. In many states, these water quality standards are widely violated without much regard. Under the 1972 Act, the EPA (or the states under delegated authority) must issue permits, referred to as NPDES permits, to all industrial and municipal facilities discharging wastes into streams. These permits, which are the primary means of directly controlling pollution from discharges into streams, state the conditions under which wastewaters may be discharged under provisions of the law, EPA industrial standards and regulations, and also the results of any planning which may have been approved. (Through a curious legal anomaly the Corps of Engineers was originally required to initiate this huge permit system under an interpretation of the 1899 Refuse Act. However, PL 92-500 put this function where it belonged, with EPA.) Detailed guidelines for industrial plant permits applicable across the country to each industry are published by EPA. The industry guidelines are the basis for negotiation between the EPA and the plant management, the final definitive treatment requirement being incorporated into a permit which specifies allowable effluent loadings or concentrations of various pollutants which may be discharged.

As of 30 June 1975, the EPA and the twenty-seven permit-granting states together had issued permits to 36,800 industrial and municipal dischargers, or 69 percent of the estimated total.[14] However, a recent court decision struck down the administrative interpretation restricting application of the permit system only to wastewater sources of a given size. The ruling seems to require permits to be issued additionally for 1.8 million animal feed lots, perhaps 100,000 stormwater point discharges, and many other agricultural and forest activities.

Even without the expanded scope of permit activities, the permit program is encountering difficulties and delays, according to a Comptroller General report.[15] In most of fifty industrial permits reviewed, the effluent limitations were not based on final industry guidelines. The industry guidelines which had been issued were under challenge by 145 lawsuits in June 1975. Nationwide adjudicatory hearings were pending in September 1975 on 450 (23 percent) of the major industrial permits. Some dischargers were not adhering to their schedules, effluent limitations, and reporting requirements. The EPA reported that almost all of the municipal permits which have been issued will have to be reissued in FY 1977, and that over half of the municipalities will not meet their specified requirements in 1977, largely due to delays in federal funding.

However laggard and defective it may be, the permit system is an indispensable part of a governmental water pollution control program. It primarily needs a sound planning approach to set the requirements which the permit should embody, and adequate systems to monitor and enforce those requirements. Thereafter, its effectiveness will depend upon great administrative effort and sufficient funding.

Nonpoint Sources and Unrecorded Pollution

The failure to consider urban runoff and nonpoint sources of pollution has been a major deficiency in water quality policy and is only recently being recognized. A study for the Council on Environmental Quality[16] indicates that after the stage of secondary waste treatment is reached, between 40 percent and 80 percent of the remaining biochemical oxygen demand (BOD) comes from sanitary sewage overflows and bypasses around the treatment plants and storm sewers. The National Commission on Water Quality staff report[17] estimates the following percentages of pollutants coming from other than point sources: BOD 37 percent, suspended solids 92 percent, phosphorous 53 percent and fecal coliforms 98 percent. This matter is discussed in detail in Chapter 4. However, it is obvious that the omission of these major sources of pollution from the official program is important. As Russell Peterson, Chairman of the Council on Environmental Quality says, "treating municipal and industrial point sources alone will not give us clean waters. We will also have to solve the problem of stormwater runoff in urban areas if we are to obtain the full benefit of our investment in water pollution control."[18]

The fifth annual report of the Council on Environmental Quality estimates that $132.4 billion will be spent for water pollution controls in the ten-year period ending in 1982. The report says that "control of nonpoint pollution is likely to become a major priority for water pollution through the early years of the next decade." According to this report, sewers served 163 million people of the total population of 210 million in 1973. Secondary treatment was provided for wastes of 104 million people.[a] However, the amount of BOD discharged by municipal treatment plants has remained constant since 1957, despite additional treatment capacity and improved plant design. The increase in pollution loading collected by the sewers, from 16.4 million pounds of BOD to 27.1 million pounds, was sufficient to offset the improvements in treatment effectiveness. This failure to attain a net reduction in pollution from treatment plant effluents is particularly disappointing because this point source pollution has been the target of the entire program to date, while pollution from urban runoff and other nonpoint sources unquestionably continues to increase with increases in population and economic development. Individuals damaged by pollution from minor sources may complain vainly for months and even years, finding federal and state officials all too busy, and county and local officials insufficiently knowledgeable, or inadequately staffed to handle specific situations. For example, a Rutgers University faculty member attempted for two years to obtain corrective action against a small industry polluting the stream above his home. Laboratory analyses showed the presence of copper in toxic concentrations up to 62 mg/1 and other heavy metals in the stream, which flows into the Millstone River, a regular source of water supply. The best result he was able to obtain was temporary alleviation of the pollution from time to time. The failure

[a]For definitions of technical terms, see Glossary.

to provide institutions with clear-cut areawide or regional responsibility for water pollution control insures that most of the smaller cases of pollution will go unchecked.

As previously noted, the 1972 Act has established water pollution control objectives independent of other water resource considerations such as water supply and flood control. This may have been desirable initially to facilitate the early start of a new program, but it is now allowing some awkward situations to develop, particularly with regard to water supply. The energy crisis has highlighted water needs, because the planning to make the United States more self-sufficient in energy has encountered water as a critical limiting resource in many situations. Former Interior Secretary R.C.B. Morton, Chairman of the U.S. Water Resources Council, even warned that water may even replace oil as America's most urgent need. While this hyperbole need not be taken literally, it does represent a needed warning. The planning for water supply in many areas is based upon diverting most, if not all, of the available low water flows of streams for consumptive purposes (which consume a large part of the water diverted, such as irrigation). This occurs not only in the arid west; a good example in New Jersey will be discussed later.

Some of the wording of the 1972 Act indicated an intent that augmentation of low flows by reservoir storage should be included in federal pollution control programs in appropriate cases, but other language in the Act has been interpreted by the EPA so as to prevent this. According to the interpretation of the law, the same degree of treatment would apparently be required whether the stream had its normal low flows, had them augmented by reservoir storage, or had them diminished by upstream diversions. In effect, the Act, as administered, appears to contemplate effluent treatment as an end in itself, to be required regardless of the quality of flow in the stream in question, either at the present time or in the future. Presumably this policy will be revised, if the alternatives are clearly exposed by planning.

Controlling Environmental Impacts

The preservation of environmental quality in our waters and adjacent lands, as explained in Chapter 2, should be the objective for our entire water pollution control system. However, the operational controls of the 1972 Act are directed towards removing pollution in all cases, regardless of whether a particular action will actually improve the environment. In other words, pollution control is a surrogate objective for improving the environment.

The EPA itself does not evaluate the environmental impacts of various proposed actions contemplated by industries, utilities or other governmental agencies, or carry on general planning to improve the environment. The U.S. Fish and Wildlife Service, the Bureau of Outdoor Recreation, and corresponding state agencies have not been expanded to play a major role in the evaluations of environmental quality now required by NEPA (The National Environmental Policy Act of 1969). Even the Council on Environmental Quality plays more of an analytical than a planning or executive role. Therefore, the protection of environmental interests, other than by the specific programs of the federal agencies, is largely left to the courts to decide.

The basic requirement for environmental impact assessments and evaluations had been laid upon the proponents of any proposed new project or program that could affect the environment. Thus, if some utility company needs to expand a power plant to meet the demand for electricity, the environmental impact evaluation must be prepared by the proponents of the project. No matter that the utility company does not have any expertise in many of the environmental aspects concerned, which are much better known by local, state and federal officials. What happens is that the utility company hires a consulting firm to prepare a study, usually voluminous, full of routine, detailed information such as geology, weather and lists of birds of that state, and extremely scanty as regards the nature of the biological and social impacts of the proposed activity, since these are very complex problems. Designated public agencies are required to review and approve the report. A public hearing is required, and since the facts are usually unclear, any determined opposition usually results in long delays. If the responsible agency approves the project, the opposition can appeal to the courts, which are poorly equipped to understand the technical issues involved. The upshot of this situation is that a small determined group can block almost any action involving environmental considerations, for a prolonged period.

This situation was dramatically illustrated when the chairman of the board of Consolidated Edison Company (Charles Luce, former Assistant Secretary of the Interior), revealed the predicament of the company in attempting to provide necessary new power plant capacity through a pumped-storage hydroelectric plant at Storm King Mountain. Both at the state and the federal level, the question of the need for a facility is decided by one agency and the environmental acceptability is the business of another agency. Decisions made are often appealed to the courts, and delays may last for years. He urged that there be one entity, whether state or regional, which would have authority to bring together all of the economic, engineering and environmental considerations and to make definite decisions.[19]

The power of the objecting group under NEPA procedures is largely negative. Objectors can suspend all action, or less often, reject the proposed action; they can block a proposed project, but they cannot indicate a satisfactory alternative.

The most recent example of major environmental conflict in the East is the 1975 decision to not support the building of the proposed Tocks Island Dam (after years of bitter controversy). This decision gave no guidance as to where the region is to obtain corresponding water supply, low flow augmentation, or flood control, but at least it finally cleared the way for studies of alternatives. It is noteworthy that the decision did not arise out of the planning activities of the federal agencies or of the Delaware River Basin Commission, but as a result of political pressure, sparked by environmental groups. In a case as important as this, the use of political channels is probably necessary and appropriate, but for lesser activities it would be useful if publicly funded agencies having real expertise were charged with evaluating environmental impacts in fields of their responsibility.

The present system, utilizing consultants paid by interested parties, has serious weaknesses. As regards environmental impacts, there is usually only extremely scanty data, which may be selected or interpreted to prove a point. Since the known technology of evaluating environmental impacts is insufficiently established to provide a firm basis for professional opinions without extensive investigation, the objectivity of environmental impact studies prepared in this manner is suspect, as is the objectivity of the original objections raised by environmental groups. The recent proposal for an environmental impact study as a preliminary to raising the New York City subway fare from thirty-five cents to fifty cents may appear initially far-fetched, but there might indeed be environmental impacts of raising the price of public transportation; for example, it might further increase dependence on the automobile. One's mind boggles at the thought of suspending New York's long overdue adjustment to the fiscal realities on account of an environmental impact study, but other activities suspended by environmental impact controversies also suffer economic or social disadvantages by long delays.

Institutional Aspects

When, in the 1930s, President Franklin Roosevelt decided to implement an exceptionally far-reaching plan for the Tennessee Valley, he created the Tennessee Valley Authority to carry out the planning, construction operation of the work as a federal undertaking. A generation later, an extensive plan for the Delaware River Basin developed by the Corps of Engineers occasioned the creation of the Delaware River Basin Commission as an implementing agency, this time in the name of four states acting jointly with the federal government. However, even larger and more complex systems of dams and reservoirs were planned and authorized for the Columbia and Missouri basins and no special implementing organization was created; the task was left to the federal agencies concerned, to be coordinated by federal interagency committees in conjunction with the states. Similarly, there has been no creation of regional institutions especially for water pollution control. The existing regional authorities which have functions related to water pollution control have been given no support or encouragement by the 1972 Act.

This situation is in sharp contrast with governmental arrangements in the Ruhr region in Germany, and more recently in England and France. (Described in Chapter 7.)

These organizations provide for regional planning, financing, monitoring, and in certain cases construction, operation and maintenance of facilities. There is no question, as in the United States, of whether or not to coordinate water pollution control actions with water supply, preservation of the environment, and water-based recreation, since the regional authority is responsible for integration of all of these aspects. There is no attempt to formulate specific effluent standards applicable nationwide, since it is acknowledged from the start that standards will vary from region to region, depending upon the plans arrived at in that region.

Naturally, procedures of other countries cannot be accepted uncritically for application in the United States, but we can well afford to take the time to consider how these countries solve certain problems and how the same problems are dealt with here, under our system of government. It is not impossible to redistribute governmental functions in the United States, if the need is great. There was a time, not so long ago, when every small community had its own little high school; however, professional opinion in favor of consolidated high schools became irresistible and previously independent school districts were consolidated, practically nationwide. The results of the areawide planning now being started under Section 208, PL 92-500 should reinforce the inferences from the NCWQ Report that new flexibility is required in our water pollution control approach. Standards of effluent treatment applied uniformly, nationwide, can no longer be accepted uncritically as a sufficient means of cleaning up our rivers, but only as one aspect of a comprehensive approach, developed by planning.

Regional Water Quality Control and Basin Commissions

Some years back, a few analysts led by Allen Kneese[20] became interested in the possibilities of regional management of water quality problems, meaning by this a comprehensive basinwide approach such as that of the Ruhr Valley of Germany. In the United States, the Delaware River Basin Commission (DRBC) was the prototype organization for regional water resources management. It was designed for comprehensive planning and implementation of both water supply and water quality, using the combined powers of federal and state governments. Upon its creation, the DRBC organized a competent staff and soon made considerable technical progress, but became largely impotent in major aspects of pollution control because of the following events: First, the 1972 Pollution Control law established federal controls and standards over water pollution which, in effect, overrode those of the DRBC. Second, in implementing the areawide waste treatment planning provisions of the 1972 Act, the most important planning for the Delaware Basin was assigned to another group, with greater funding than could be undertaken by the DRBC. (Under a little-noticed provision of the 1972 Act, areawide water quality planning must be assigned to agencies legally competent to conduct land-use planning; but they are not required to have any experience or qualifications in water quality planning.) The states of the region might have united in 1972 to defend the DRBC, but in the excitement of that year it would have been a bold politician who would have risked incurring the wrath of environmentalists by holding out for major modification of the 1972 Act, which passed the Senate without a dissenting vote. The four states concerned allowed the DRBC to be hamstrung without effective opposition. These changes severely impaired the usefulness of the planning and pollution control functions of the DRBC and other existing basin commissions. It should also be noted that most of the commissions did not have pollution control functions, except for aspects inherent in comprehensive water resources planning.

There appears to be a complete lack of priority for the regional planning of water pollution control on interstate rivers, except for a few cases where portions of them happen to lie in a metropolitan area. This is a serious anomaly, because the interstate rivers provide the best reason for putting the federal government into the water pollution field in the first place. A generation ago, any attempt by the federal government to regulate water pollution in rivers other than interstate would unquestionably have been ruled unconstitutional by the Supreme Court.

The only recent federal attempt to develop pollution control plans for the Delaware Basin as a whole was included in a framework plan for the North Atlantic region, which was prepared by federal agencies including the EPA (and its predecessors). The pollution control provisions were totally inadequate; only one proposal for pollution control was specified, which was advanced as being equally applicable to maximizing both the environmental objective and the economic objective. The plan was to impose 90 percent removal of BOD and 95 percent of suspended solids throughout the basin, at a cost for the next fifty years of $14 billion. After the expenditure of this vast sum of money for treatment it was estimated that the pollution remaining in the estuary would have *increased* sevenfold. The report did not comment as to the probable results of such a pollution increase upon any already overburdened estuary, did not suggest any alternatives, and did not return to determine what had been done wrong. In short, this huge report, published in twenty-six volumes, which took eight years and over eight million dollars of federal money to accomplish, ended as a complete absurdity as far as pollution is concerned.[21] The public and the media took no notice of this astonishing result, apparently having previously arrived at the conclusion that nothing would come of the planning activity.

Results of framework planning were similarly disappointing for the Hudson River. Although the Water Resources Council has adopted procedures that should improve results of comprehensive planning in the future, the lack of serious pollution control planning, including consideration of both economic and environmental objectives, remains a severe deficiency in the general water planning picture.

Intrastate Planning

Even for rivers wholly within a state it is not simple to prepare an adequate scheme of water and wastewater management. Within the state of New Jersey, the Passaic River is the most obvious intrastate problem area. The basin study area includes a population of 3,800,000, including Newark, Jersey City, and Patterson-Clifton-Passaic. The population density is a remarkable 3,400 per square mile, in spite of large undeveloped areas in the upper watersheds. It includes all of four counties, parts of seven others and 181 municipalities, plus some headwaters areas in New York State.

The water problems of the basin include an extremely polluted condition of the lower river and the estuary, a contant growth of population and industry spreading ever further upstream, with consequently increased pollution, and the necessity of maintaining the upper part of the basin for water supply. As of 1972 the basin had 149 wastewater treatment plants, of which 17 still had primary treatment. These include eight of the nine largest plants, specifically the mammoth Passaic Valley Sewage Commission plant of 250 mgd (million gallons per day).

Water supply withdrawals by ten water companies totaled 563 mgd in 1971, which in the case of five of the companies exceeded the safe yield of water at the point of withdrawal, computed on the basis of the drought of the sixties.

Two major public water-related institutions in the basin struggle against the mounting pressures of the situation. The Passaic Valley Sewage Commission (PVSC) operates a trunk sewer system which collects most of the effluents from the lower part of the basin and treats them in a central plant, and it takes action against observed unauthorized pollution. Upstream, the Passaic Valley Water Commission, which treats its river water at its water supply diversion, initiates action by any means available against illegal pollution upstream.

The state has let a contract for a consultant to prepare a water quality study of the basin and, a second water supply study will be made incidental to a statewide water supply plan, which is about to be initiated through another consultant contract. The upgrading of the huge PVSC treatment plant which has already been initiated (after federal pressure of some years) will show that treatment of other point sources should be improved. As early as 1969, research in the upper basin revealed that the greater part of the pollution did not come from known effluents but from unrecorded sources.[22] More recent research in this area has shown that heavy loads of organic pollution, nutrients and heavy metals originate in the urban areas.

The Passaic River Basin is better provided with institutions than most, because it has two vigorous public agencies which are attempting to control indiscriminate pollution. However, further planning required for this area should not be done by two independent consultant contracts, one for water supply and one for water pollution. It is obvious that diversion of water supplies has been pushed towards the maximum, with complete disregard for water quality of the river. What is involved is a basic conflict of interests, in which either development and growth in the upper basin must be limited or watersheds (and many environmental advantages) must be abandoned. As far as the lower basin is concerned, the evaluation of urban runoff pollution should be the highest priority for study, since advanced waste treatment of the point sources alone will obviously be insufficient to achieve water quality standards and may be uneconomical.

It would probably be desirable to have water and sewerage in the Passaic River Basin controlled and monitored by a single institution. If not, a more complete delineation of responsibilities and authorities would be desirable. The preparation of consultants' plans and review by the state may provide some correct indications of action to be taken, but the means for carrying out such action will remain to be determined, especially with regard to land use control.

This situation is not unique, but characterizes many rivers on the eastern seaboard, where population is heaviest towards the coast and water supply comes mainly from headwaters. What is unusual is that in this case the potential conflict has progressed as far as it has.

National Commission on Water Quality

The 1972 Pollution Control Act itself contemplated and authorized a reconsideration of its approach. It called for the appointment of a National Commission on Water Quality, which was charged with examining the effect of implementing the goals of the 1972 Act and considering whether any change in those provisions would be desirable. In November 1975, the commission issued a "Staff Draft Report" that gives the results of the many studies the commission has made.[23] This report has a great deal of valuable information which is referred to in this book as "NCWQ Staff" findings.

The NCWQ Staff Draft Report estimated the costs of achieving nationally a number of the goals specified in the 1972 Act (see Table 1-1).

In contrast to these huge estimates, a key congressional aide testified at the 1976 Water Pollution Control governmental affairs seminar that probable appropriation levels for this program might be $5 billion for FY 1977 and $6 billion each year for the two following years.[24] It is apparent that the program as now visualized is far too costly to be funded.

The operation and maintenance costs of these enlarged systems would also

Table 1-1
Estimated Costs for Implementing P.L. 92-500

Costs	Billion $
Costs of achieving 1977 treatment goals for publicly owned plants, Federal outlay only (75%) (For 1983 goals practically the same cost)	118.5
Non-federal costs of achieving 1977 goals for publicly owned plants would presumably be 1/3 of federal cost or	39.5
Correction of combined sewer overflows	79.6
Treatment of urban runoff only to the extent of suspended solids removal and disinfection	199.0
Capital costs for industry to meet 1977 goals	44.3
Added costs for industry to meet 1983 goals	30.6
Total	511.5

increase. The operation and maintenance costs for publicly owned plants (exclusive of combined sewers and stormwater treatment) would be $2 billion more annually by 1990 than in 1973. Industrial operations and maintenance costs would increase by $7.7 billion annually to meet 1977 goals, exclusive of costs of pretreatment for industries discharging into publicly owned systems. To meet 1983 goals, operations and maintenance costs to industry would increase an additional $6.2 to $6.9 billion annually.

It should be noted that the Commission did not attempt to estimate the costs of meeting the stated national objective of eliminating all discharge of waste materials by 1985. This objective of the legislation is so impracticable that no one now takes it seriously.

The NCWQ Report also estimated the probable effects on water quality of applying the provisions of the Act, but without the control of urban runoff. The effectiveness would be greatest in improving dissolved oxygen in rivers. During low flow conditions, 380 miles of the rivers studied now have unsatisfactory DO conditions; the 1983 program would reduce this to 23 miles. However, this evaluation does not include consideration of nonpoint source pollution, which would reduce the indicated effectiveness. It was concluded that fifteen out of thirty-eight study sites would require more stringent treatment requirements than application of the 1983 standards in order to meet satisfactory conditions.

The study found that the decrease in total phosphorous due to the program would be only marginally significant in terms of influencing eutrophication problems. Only about one-third of the sites studied would show significant reductions in suspended solids and turbidity. Projections of coliform bacteria show that there would be considerable improvement due to treatment programs, but that over 70 percent of the sites would still have periodic coliform bacterial contamination because of nonpoint source loadings.

Data are not available as to reduction of toxics such as heavy metals by the projected program, and at most sites data were not available to quantify the responses of fish populations to abatement of point source pollution. Generalized studies indicate that a major improvement in conditions for fish would probably occur in the northeastern United States, but there would be little effect elsewhere. Somewhat greater effects are expected for shellfish at estuaries, but the analysis is lacking the information required to be more specific.

An estimate was made of the economic benefits of the program, although admittedly such estimates can only be very approximate.

The cumulative benefits over the decade 1975-1985 are estimated as indicated in Table 1-2. It is pointed out that some of these benefits (averaging $3.6 billion annually) would continue indefinitely thereafter. What was not emphasized was that operations and maintenance costs totaling $16 billion annually would also continue indefinitely thereafter, as well as the interest on the bonds to finance the many billions of dollars of investment. In short, there was no attempt to show that the benefits from the program exceed the costs, and the data produced strongly suggest that they do not.

In summary, the NCWQ Report clearly indicates that the presently programmed implementation of the 1983 *treatment* goals (for waste treatment plants) would be only partially effective, and that treatment of urban runoff and combined sewers would be required in order to reach the 1983 *water quality* goals. This would cost a total of some $511 million dollars by 1983, a sum far greater than is practicable.

The final recommendation of the National Commission on Water Quality[25] at least partially followed the findings of the staff and the consultants, even though these findings themselves were not formally endorsed. The 1977 goal for treatment plant compliance is recommended to be maintained, but flexibility is to be allowed for EPA to grant extensions and even waivers. The most important provision recommended is the postponement for five to ten years of application of the 1983 standards of "best available" treatment technology, while maintain-

Table 1-2
Estimated Benefits for Implementing P.L. 92-500

Benefits	Cumulative Value $ billion
Increased property values	.486
Reopening of beaches	6.4
Marine resource use	21.3
Fresh water fishing	4.0
Other boating	4.2
Total	36.4

ing the 1983 goal of fishable and swimmable waters. Although it is apparent from the magnitude of the NCWQ cost estimates that full achievement of the water quality goals by 1983 is utterly impracticable, the recommendation at least would direct the program towards the improvement of water quality, rather than towards the application of advanced treatment technology for its own sake.

The 1985 goal of eliminating all discharge of pollutants is to be "redefined" in other words, abandoned as quietly as possible. The achievement of 1983 goals for irrigated agriculture is to be sought with "flexibility." In other words, the 1972 Act in this respect made an unwise provision, which is to be reconsidered and modified. The net effect of these recommendations, if adopted, would be to redirect the decision making procedures of EPA towards a planning process rather than towards attempts to apply nationally uniform standards and criteria, which have been shown to be intolerably expensive and also inefficient in achieving the objectives they purported to serve. No one knows when and how the Congress will act on this matter, but it appears that rational planning processes which have so far been given only lip service should now become a more influential means of allocating resources in the water pollution field. However, the planning methods have not been well-developed, and many of the principles are obscure. The chapters that follow are devoted to this subject.

2 The Environmental Quality Objective

Let us a little permit Nature to take her own way; she understands her own affairs better than we.

Montaigne, Essays

Introduction

The question of how to determine the social value of environmental goals is one that plagues all serious analysts of resource management. Although in 1960 it was almost impossible to make citizens aware of the problem, in 1970 the Gallup poll found that citizens considered environmental pollution to be the second most important problem of the decade.[1] The new emphasis on environmental aspects has discredited most of the planning approaches of the past without offering any operationally valid alternatives. It is essential that environmental concepts be understood thoroughly if water quality planning is to be effective, and that operationally practicable technologies of evaluating environmental quality be developed.

One reason it is so hard to state our ultimate objectives is the increasing preoccupation in American life with novelty for its own sake. We have been cautioned that in some circles change and newness are on their way to becoming an obsession, as if more defined identity were only too much conformity.[2] Therefore, in exploring the new ideas, we must be careful not to lose touch with fundamental values.

The Environmental Mystique

Environmental groups proliferated rapidly from 1969 to 1972. Gerlach[3] says that the diverse groups concerned could be arranged on a continuum from the established and routinized to the new and radical. Common ideological concepts of the more radical spokesmen are: the doomsday theme, general guilt-sharing for environmental degradation, the concept of finite earth resources leading to a zero sum game, the closed system of spaceship earth, the need for recycling, the need to control or limit growth, ecosystem and interdependence concepts, the need for significant change, and the thought of younger members that system change means lifestyle change. From another viewpoint the great bulk of the more than abundant environmental literature may be classified into two groups representing the preservational and the conservational aspects of the environment movement. Two other groups are the traditional planners, who oppose the entire movement, and the eco-activists, who believe that adequate environmental solutions are not possible within the context of existing societal structure and value.

The preservational aspect is the more easily defined, being dedicated to preserving nature in its "original" state, untouched by man or at least substantially untouched. In this view man is considered to be no more important than any other species, and the preservationist usually tried to minimize man's presence in preserved areas as far as practicable, primarily for ideological reasons. An excellent picture of preservationists' beliefs and approaches was contained in a series of *New Yorker* articles in 1971.[4] The pure view of preservationists is that all natural conditions (i.e., nonmanmade) are good in themselves and that it is a major tragedy when some species becomes extinct or some type of natural ecosystem ceases to exist, no matter how distantly related to man's interest and activities it may be. In this view, any altered environment is inherently less desirable than any unaltered environment, no matter if the altered environment may provide much more abundantly for man's needs, including recreational and living space. The preservationist's viewpoint is inherently antitechnological, viewing the construction and industrial activities of western civilization as eroding the quality of our environment.

The conservational aspect represents more complex and diffuse objectives, inherited in part from the national conservation movements of the past, which emphasized wise and restrained use of our natural resources in the interest of national welfare and urged consideration in planning environmental objectives, including those related to man's desires and activities. Although the two groups often coincide in their practical objectives, their conceptual bases are quite different.

In fact, although it is not generally realized, the preservationist and the conservationist are in some respects irreconcilable. In our rapidly developing metropolitan areas, the provision of living and recreation space for the urban populations, and the pollution that their activities generate, are the principal forces that are taking or spoiling the natural habitat of species other than man. It is increasingly difficult to reconcile man's needs with the preservation of a natural environment. Although wise conservational policies and advanced technology can ease and at times postpone the conflicts, there is a basic conceptual difficulty that must be faced in the interests of rational thinking on the subject.

The concept of the *environment*, when used as a slogan, often succeeds in rallying conservationists and preservationists in a common attack on some proposal for construction, but such unity is mainly on a political level and the underlying problem of basic aims remains. Overt disagreement is often avoided by the varied meaning given to the word "environment." The term is used by some specialists to refer only to nature or to ecosystems,[5] whereas planners and architects often broaden the definition to include anything physical and tangible exterior to man himself. In the analysis presented here, environmental objectives or attributes referring only to natural species will be referred to as *ecological*, the term *environment* being used in a broader sense to include those related to recreational and living environment. Besides the obvious water-related activities such as fishing, swimming, and picnicking, these values include such attendant pleasures as courting, bird watching, nature photography, backpack camping, and cross-country skiing.

Worldwide Environmental Problems (The Doomsday Approach)

It has become popular to suggest that in the not too distant future, our worldwide civilization will come to an end in the welter of its own pollution, or suffer other worldwide disasters consequential to a lack of environmental foresight. The Club of Rome Study[6] was the most famous of these views; it was based upon extrapolation of geometrically increasing statistical trends. An example frequently used in demonstrating a potential global environmental problem is the "hothouse effect," related to man's impact on the chemical composition of his atmosphere. Changes in proportions of carbon dioxide, of particulate matter and of waste heat in the atmosphere are thought to be real threats to the stability of worldwide atmosphere and climate. All such apocalyptic predictions are successors to the famous Malthusian predictions, made in 1797, which envisioned the world soon entering a crisis in which the growing population would outrun its food supply. As regards global environmental problems, perhaps the most useful concept is that of Adlai Stevenson, quoted by Dubos[7] which envisages the spaceship earth depending upon its vulnerable supplies of air, water and soil, with all aspects of its environment interdependent. Inevitably, the exhaustion of earth's natural resources will require that its future economy be based upon sound ecological principles and careful residuals management.

Although certain worldwide trends undoubtedly constitute a long-run ecological danger, and should be studied, by far the greatest environmental deterioration and the most immediate problems are found in the world's great cities and industrially developed regions. It is to these immediate regional problems and not to the question of the survival of spaceship earth that this study gives primary attention.

Crowding and Stress

One physical aspect of environmental quality appears to run through all theories—namely, that of space for man himself. Dubos feels strongly that the adverse social effects of overpopulation or the "... crowding (of humans in cities) may lead to types of social behavior that recall the social unawareness observed in overcrowded rodent populations." He states that crowding will lead to spreading of urban and suburban blight, traffic jams, water shortages and all forms of environmental pollution. In the long run in such environmental conditions, natural selection will tend to eliminate from the human race those who maintain philosophical values such as literary, artistic, and religious ideas.[8]

Specific environmental stresses may cause not only psychological but also physiological damage.[9] among such stresses are the congestion, crowding and confusion of the inner city. Rates of mental illness, suicide and drug addiction may be used to indicate the presence of anomie and the inability of individuals to adjust to urban pressures.[10] One study of 2000 persons in midtown Manhattan showed a very high level of neurotic and psychotic individuals.[11]

Increased population densities, noise levels, congestion and social problems cause so much tension in the (urban) American that the major motivation of many vocationists is to escape. Recreation areas should provide opportunities for disengagement, locomotion, variety, change, diversion, isolation and withdrawal. The surveys of public opinion and attitudes summarized elsewhere in this report show how prevalent the motivation is to escape the pressures of society to a spacious natural environment. Frederick Olmstead over a century ago justified the value of parks for the re-creation of man as well as recreation.[12]

As would be expected of him, Thoreau in 1851 expounded the extreme virtue of withdrawal from society: "In wilderness is the preservation of the world." Few individuals feel such thoughts so strongly. There are also some very opposing views and feelings on the subject: many people enjoy a crowded urban environment. Even regarding recreation choices, the spatial occupation of the Jones Beach and Fire Island bathing areas on Long Island clearly shows that the vast majority of the hundreds of thousands of persons present must either prefer the more crowded areas, or are so little attracted by privacy, quiet and additional space that they are unwilling to walk a few hundred yards further to obtain it.

Between different national cultures, the need for privacy or for intimacy in social contact may vary greatly with regard to sight, sound and olfactory perception so that "the kind of private and public spaces that should be created for people in towns and cities depends upon their position on the involvement scale."[13,14] However, even though people under some circumstances may welcome the warmth and excitement that accompany crowds, most of the visitors to parks and wilderness prefer that natural environments be sparsely occupied. The environmental value of any area depends greatly upon the space available.

In considering these ideas, we have as a guide the various biological experiments that show the ultimate population limits of a single species growing in a limited environment. When flour beetles, mice or other laboratory animals are allowed to multiply within a given limited space or with a limited food supply, disasters may occur. Either the community generates waste products such that it exterminates itself, or the populations cease to breed new young, or various sorts of general antisocial activity may start. Paul Errington's fascinating work on marsh ecologies[15] describes in realistic detail the tensions and conflicts in muskrat populations exposed to overcrowding. The human race is not yet so numerous on earth that catastrophe appears imminent, but various stresses are beginning to occur due to the vastly changed environment in which urban man, in particular, now finds himself. The evidence lends strong support to the widespread perception that environmental quality declines as urban scale increases. To the majority of people, behavior in large cities appears colder, more impersonal and less polite, and life is liable to be more frenetic and irritating.[16] The noise, congestion, dirt, and polluted air of our cities, and the relative inaccessibility of open space with trees, grass, streams, birds and fish provide an environment that is in some respects undesirable. The opening of fire hydrants in ghetto streets on hot afternoons is a rough expedient to improve an environmental condition felt to be intolerably bad. This much is obvious.

A recent theory of urban stress[17] implicates certain environmental conditions as contributors to the stress syndrome, especially poverty in combination with social disorganization and overcrowding. Where such conditions are accompanied by poor adjustment on the part of individuals, the results are pathological: violence, death and insanity.

The incidence of crimes of violence is said to be correlated with city size (although according to some recent reports the suburbs are beginning to be similarly affected). However, all individuals are not affected similarly, and a stimulus does not constitute a stress unless perceived to be so. In summary, it is extremely difficult to define just what constitutes spatial quality of the urban environment in terms that will be generally acceptable. It is even more difficult to measure its value to society.

There is no common agreement as to the nature of ultimate aims in an urban society. The relatively high levels of consumption and of energy availability in the United States do not prevent our having urban problems more severe than those of Europe. In fact, the large quantity of goods and services produced and consumed in our society may be regarded as evidence of priority for secondary activities, which may unduly divert our resources and efforts from the enrichment of life in other ways.[18]

Man's Effect Upon His Environment

In one ecological sense man is but one of the species inhabiting the earth, but this knowledge does not get us very far. All natural species compete for survival in the Darwinian sense and tend to expand to the limits of their capabilities, either by preying upon other species, by depriving them of food supply, or in some cases by physical crowding or chemical inhibitors. Man has so successfully exploited the earth that other species are rapidly being reduced in population and range, or exterminated altogether, a fact that poses two main questions, the first of which may be found to be a subset of the second: First, is it really the desire of mankind to advance to the virtual elimination of other sizeable species (except for tame animals and garden plants) and, second, at that point does the crowding and pollution produced by man become inimical to his own further development?

Man's activities affect ecosystems mainly in three ways: by the physical occupation of habitat, by pollution, and by predation, such as over-fishing. When the effect is directly harmful to commercially valuable species, such as massive fish kills of trout or flounder, the activity is readily understood to be undesirable. However, the extent to which natural species are dependent upon each other as well as with various attributes of the environment, and the complex ways in which they affect man's interests, are only recently becoming widely understood.

Rachel Carson, in her best seller, *Silent Spring*, published in 1962, describes the delicate networks of interdependency between plants and animals and their vulnerability to man's interference. The drifting plant life, the insignificant microscopic plankton, the infinitely small, fragile-as-glass larvae of oysters and clams, of which the industrial water user is not even aware, are all potential targets of the wastes of modern society.[19] The myriads of plankton form the base of the food chain of the ecosystem and they are highly vulnerable to toxic substances from water pollution.

Since plankton are not directly important to the farmer spraying the crops or to the manager of the waste treatment plant, they will not necessarily seek alternatives which would preserve these apparently insignificant species. They are operating rationally within the profit-maximizing framework of the free market system; to do otherwise would be irrational. Therefore, for environmental protection we must look mainly to government. Fortunately, during the last few years there has been a great change in governmental attitudes toward ecosystem preservation, and this is due almost entirely to the environmental movement.

Interacting complexes of living organisms and their biotopes (physical subsystems) are called ecosystems.[20] The life forms that inhabit biotopes are considered as inseparable from the biotope. We are just beginning to understand the functional attributes of ecosystems and of man's role as manipulator. The capacity for biological communities to function within ecosystems and the vulnerability of community function to environmental disturbance depend upon functional relationships that allow population interaction. The capacity of species populations to survive major changes depends on mutation and gene recombination rates in relation to the rapidity of change in the environment.

The fundamental characteristics shared by life make all forms of life on earth susceptible to a wide range of environmental inhibitors. Species vary in their tolerance to environmental stress due to a wide variety of protective mechanisms. In general, populations are restricted to a finite range of stress intensities beyond which they cannot exist. A stress in this case can be the presence or absence of a required nutrient as well as the presence of a toxic material or an adverse temperature. Because of the great biotic variation between species, the tolerance of a community to environmental stress is greater than the tolerance of any single species.[21] There are many ways in which biosystems may be affected by pollutants or by man's conversion of habitat to other purposes. These relationships are only beginning to be understood in detail, however, in general, the harmful effects of pollution upon species of interest to man is quite clear. Even though large numbers of other biota such as masses of algae and sludge worms, can grow in the polluted waters, the interest of the stream to man is greatly reduced.

The General Environmental Problem

In attempting such a complex task as evaluation of environmental quality, it would be unwise to rely upon popular concepts or approaches derived from single intellectual disciplines, since important aspects would be sure to be overlooked. The disciplines of engineering and economics were most concerned with this problem initially, but as the theoretical constructs of these disciplines fail to encompass some of the most meaningful relationships, much broader systems of thought are required. The inadequacy of all quantitative approaches to date is due to the lack of development of basic philosophic concepts applicable to the environmental problem. The empirical approaches now being used generally lack any rational basis. Nothing less than the relationship of man to nature is involved. We must either accept or reject ideas advanced from political, social, psychological, and anthropological sources. We have been urged to reassert humanistic values, to renew the approaches of rationality, and to seek the restoration of ethical criteria in human enterprise.[22] Such suggestions may appear strange and unfamiliar to engineers and planners, but they are not necessarily irrelevant. Poetic and religious concepts may be meaningful; but before we discuss methodology, it is necessary to understand the problem thoroughly. It is best to open our minds to a wide variety of approaches, and this is done in this chapter.

Some of the irreconcilable views discussed in this chapter originate, at least, in part in cultural interests of particular groups of society. The preservationist's view, the various aesthetic views, and the religious view are in part rationalizations of a deeply felt common feeling and in part expressions of particular interest or cultures. On the other hand, many of the eco-activists' views seem to utilize a great common feeling of mankind to advance a person's particular view, in this case political.

Philosophic Approaches

It must be emphasized at the outset that the problem of value of the environment is a subset of the general philosophical problem of the ultimate significance of human values, for which, given the diversity of the metaphysical and cultural approaches of mankind, it would be futile to anticipate a precise solution. Words such as *value* and *welfare* are used as convenient symbols for the unknowable, or rather the imperfectly knowable, and their precise quantitative application to practical problems is not to be expected. What we are searching for is "criteria that could serve as conceptually and operationally meaningful proxies for the fictional constraint of optimizing welfare."[23]

Humanistic objections to traditional approaches, emphasizing material aspects, have been perhaps best expressed by Rene Dubos.[24,25] According to Dubos, "In many respects modern man is like a wild animal spending his life in a zoo. Like the animal he is fed abundantly and protected from inclemencies, but deprived of the natural stimuli essential for many functions of his body and his mind." Dubos feels that man acquired his basic traits while living in intimate relationship with other living things, and that "the biological and cultural heritage of the Paleolithic hunters, Neolithic farmers and subsequent civilizations is indelibly incorporated in all subsequent activities of mankind. This is the historical determinism of social life." Dubos goes as far as to refer to "the genetic record of past experiences"; but he later explains, a little awkwardly, that he means only that natural selection has favored those genetic characteristics adapted to the natural environment. Genes constitute potentialities that become reality only under the shaping influence of stimuli from the environment, especially during the formative stages of early life.

It is worth noting at this point the similarity of this view with that of the analytical psychologist C.G. Jung, who emphasized the primordial image or archetype, which corresponds to the disposition inherited from ancestors or the collective unconscious. "The archetype is manifested principally in the fact that it determines human nature unconsciously but in accordance with laws and independent of the experience of the individual. When the unconscious content is perceived, it confronts consciousness in the symbolic form of an image."[26]

Although, like Dubos, we must stop short of accepting specific attitudes toward the environment (such as those of the Sierra Club) as being a part of the common genetic basis of mankind, it certainly seems plausible that there are certain genetic predispositions that are so powerfully reinforced by cultural influences during our early formative years as to be independent of our later experience (if not of all experience). For example, the excited interest with which young children regard small living creatures must certainly be of genetic origin. To quote Dubos again, "Whatever science may have to say about the fundamental processes and constituents of the natural world, we regard Nature habitually and respond to it with our whole physical and emotional being. Deep in our hearts we still personalize natural forces and for this reason experience guilt at their descecration. The manifestations of Nature are identified with unchangeable needs of human life, and are charged with primeval emotions because man is still of the earth earthy."[27]

Dubos explicitly rejects the preservationist's desire to retain or restore a primeval environment, which he considers would be not necessarily either desirable or meaningful. He feels that the developmental characteristic of Western civilization is too basic to be changed, and that conservation implies a creative interplay between man and animals, plants and other aspects of nature, as well as between man and his fellows. He feels that a new kind of knowledge is needed to study man's functioning as a whole, including all of the complex relationships involved with an ecological attitude very different from that current in biology. He states that the emphasis on humanistic criteria is not a retreat from science, but rather points to the need for enlargement and rededication of the scientific enterprise. The main current of Dubos' thought is to emphasize the importance of nonmaterial values in our relationship to nature, but he does not try to define such values very closely.

The view of nature inherent in the worldwide Boy Scout movement and experienced in summer camps for children is consistent with Dubos' thought. There is no doubt that children of diverse cultural backgrounds, many of whom are too young to understand complex teachings even if they had been exposed to them, love to run about, to splash in water, to play and to live in a natural environment if the weather and climatic conditions are favorable. They show an immediate interest in all living creatures. These interests are apparently fundamental and deeply felt in children, and seem to arise from sources more basic than teachings of even cultural influences.

Religion and Spiritual Aspects

From early times, man's survival depended upon bending nature to his will, as reflected in the Old Testament admonition to "replenish the earth and subdue it, and have dominion over . . . every living thing."[28] In the famous controversy over damming the Hetch Hetchy Valley in 1915, religious thoughts were invoked freely, the would-be builders being referred to as temple destroyers, and appeals being made to "the God of the Mountains."(In contrast, the conservationists' views were ridiculed as "sentimental dreams and aesthetic visions," and the conservationists themselves were called "mushy esthetes.")[29] In a recent survey, many preservationists were found to agree with the thought that "Nature, Man and God are all one," which is a central tenet of pantheism.[30]

There is no doubt that a good many people are animated by religious sentiment upon regarding some magnificent natural scene or some impressive grove of trees, such as the Druids may have used. However, it seems impossible to separate such religious thoughts and feelings from the general feeling of the importance of an attractive natural environment.

It has always been assumed that pollution or other wanton destruction of natural resources is anti-Christian, however, the opposite viewpoint has also been advocated. According to this viewpoint, the pollution of the modern world is the result of ancient Biblical texts telling man that the earth is intended for his use. The American Indians and St. Francis are praised for having believed that plants and animals are inhabited by spirits [or have souls]. It is argued that a new religious viewpoint is needed to correct the alleged anthropocentric tendency of the Jewish and Christian religions.[31,32] It hardly seems probable that the sophisticated advocates of these new ideas are really motivated by religious feeling and it is hard to excuse the flagrant misrepresentation of established religious beliefs. These articles are obviously tendentious and should not be taken seriously.

In 1948 in Montana, a violent reaction was provoked by a proposal to build a large power and flood control reservoir in Paradise Valley on the Clark Fork River. The arguments against the project were emotional. The closing peroration of the final orator was: "You engineers are trying to flood out the choicest part of God's green footstool!" Presumably that man did not believe that God sits on a Montana mountain with his feet in the valley, but he was trying to say something symbolically, as were many others present. It was obvious that construction of that dam would be regarded as desecration of something held sacred, and the message came across loud and clear.

The wilderness must be regarded not only as a physical thing but as an attitude towards an aggregation of trees, rocks, soil and water which renders it indivisibly valuable. What society takes to be a desirable natural environment is one. As Santayana says[33] "There is no value apart from some appreciation of it and no good apart from some preference of it before its absence or its opposite. In appreciation, in preference, lies the root and essence of all excellence."

The symbolic and social meanings of environment are important. The quality of the particular environment is defined by attitudes, establishing its significance and position in the context of society. The overwhelmingly abundant wild forests of the past were not highly regarded, they were dangerous, and it was with a sense of relief and achievement that the settlers cut them down. A *summum bonum* of preserving trees cannot be accepted as part of an ethic of social justice. New proxy or substitute environments serving human needs can be created to serve the needs of society.[34]

Economic Approaches to Environmental Value

Environmental matters historically have been considered of only peripheral interest to economists. The relevant early literature consists mainly of work on value theory. Other related economic analyses include the general theory of welfare economics, and some more recent studies relate to outdoor recreation. More recently, a few serious efforts have been made to develop parts of an economic theory of water pollution, which is discussed in Chapter 6. There have also been some ambitious efforts at modeling economic relationships, particularly by input-output analysis. Much of the economic literature related to environmental matters is rather dismaying to the general reader, or to the planner, since it constitutes an elaborate theoretical framework that often seems to deal with shadows and collateral issues rather than with reality. In some respects, considerable progress has been made, but at best the approach is very uneven.

The misuse of the traditional concept of economic value by planners and analysts in benefit/cost analysis really sets the stage for the environmental revolt. Since environmental values are largely intangible and not readily measurable in terms of monetary units or "willingness to pay," they are generally neglected. Largely because of this difficulty, the benefit-cost approach became generally discredited. Some analysts came to feel that in order to settle important problems of optimization, all intangibles must be evaluated quantitatively, no matter how approximately. However, Eckstein feels that the absolute measure of intangible benefits and costs "is an impossibility because of the arbitrariness of definitions, the complexity of some of the costs and benefits, and the requirements of forecasting."[35] Many other economists support this idea, stating that some intangible values will never be determined. In planners' terminology, it is considered conceptually impossible to have a single theoretically correct objective function. However, intangible values must be considered. As Leopold has said,[36] "To achieve a better and a fuller life it seems to me necessary that we look beyond the temple idol of monetary evaluation.... It would seem to me less esoteric and far more honest to decide that certain elements are necessary for better and fuller lives without dreaming up ways to put meaningless dollar values on them; then to weight the other elements having monetary value by uniform and objective procedures, so that alternative uses of the resource can be compared, and choices among the alternatives made on sound monetary grounds."

However, considerable progress has been made in evaluating benefits once considered intangible, through advances in methods of analysis (e.g., the value of outdoor recreation resources). In other cases, suggestions for quantification have been made that certainly seem to warrant further investigation.

Perhaps one of the most important new concepts to emerge recently, relative to evaluation of scarce natural resources, is from the work of Boulding. He envisaged the traditional value system as based on the "cowboy economy," the concept of "illimitable plains," where "consumption is regarded as a good thing and production likewise; and the success of the economy is measured by the amount of the through-put from the 'factors of production'..." In its place, he focused on a "spaceman economy," in which "the success of the economy is not production and consumption, but the nature, extent, quality, and complexity of the total capital stock, including in this the state of the human bodies and minds included in the system."[37] Boulding's "spaceman economy" concept brought dramatic attention to a deteriorating environment and the limited abundance of our resources. However, most economists still conceive of value as determined by willingness to pay.

Welfare Economics and Utility

Economic theorists realized some years ago that, in principle, there was an objection to adding up the utilities of various persons, no matter which definition of utility was used. If a numerical measure of utility representing willingness to pay was selected, the aggregation of utilities of different persons was considered to be improper because the marginal utility of money might be much higher for a poor person than for a wealthy person. Therefore, such utilities could be measured only ordinally, not cardinally.

To avoid such difficulties and obtain a more rigorous approach to economic analysis, a special discipline came into being, called welfare economics, based upon a concept referred to as Pareto optimality. This complex approach and intellectual blind alley is explained in Appendix B. It has long since been demonstrated that strict welfare economics approaches cannot be applied to practical problems involving large numbers of people without grossly violating the assumptions upon which the theory was based. However, reputable economists continue to pay lip service to this approach. The continued use of the term "Pareto optimality" in current economic discussions apparently represents either a nostalgic yearning after the conceptual purity that pure welfare economics represented or an avoidance of specifying the utility approach actually used, since no rigorous theory exists to support it. In either case, the use of the term should be considered to be intellectual window-dressing, not to be taken seriously.

In fact, economists in the field of public policy must add utilities of different persons if their approaches are to be relevant, just as do other analysts; and there is no reason they should not do so if proper account is taken of the economic status of different classes (see Note, Appendix B). The term "welfare economics" is now commonly used in the general sense of being concerned with general welfare as opposed to strictly market values, and does not imply the use of the approaches of Pareto optimality.

Evaluating Outdoor Recreation

All methods of measuring participation in outdoor recreation suggest an increasing national interest. The number of visits to national parks and national forests has increased between eight and ten percent annually. Recreational use of man-made lakes and national wildlife refuges has climbed even faster, as have the numbers of fishing licenses, boats, and other associated equipment. For this reason the attention to evaluating recreational resources is well directed. If water quality were improved generally, it would be expected that major recreational benefits would be attributed to this expenditure.

Although only a few years ago outdoor recreational benefits were regarded as intangible, few economists of today would agree. There is a general consensus that the economic value can be estimated based upon (imputed) willingness to pay. Profiting from earlier work by others, Clawson in 1959 finally outlined a practical approach to evaluating outdoor recreation,[38] using inferences from travel costs and time expended by proportions of the population traveling to the site from different distances in order to construct a demand curve. This approach has been extended and modified by others.[39,40,41,42,43,44] There are certain inherent disadvantages in this approach, but it is generally accepted as being capable of providing at least an approximation of public "willingness to pay" for the recreational benefit in question. Although each of the methods has disadvantages and further research is needed, there does not appear to be any barrier to obtaining reasonable approximations of the economic value of recreation.

Other economic approaches have been suggested as a means of evaluating recreational values. One of them is the gross expenditure method, which attempts to value recreation (or other nonmarket goods) on the basis of the total amount spent by the user to obtain it. This method has been used largely because it is rather easy to obtain data; also, it indicates high benefits from the proposed improvement, wich are reassuring to its protagonists. This method is relevant to the commercial interests of local concerns and chambers of commerce, but is totally inappropriate for the purpose of national decision making.[45,46,47]

Another more meritorious, although partial, approach to evaluating recreation is by reference to changes in the value of real property when recreational opportunities are improved (or reduced in attractiveness). The use of such changes in value presents problems in that possible consumer surplus (the value of something to the individual over and above the amount of money paid for it) may not be taken into account.[48] Also the benefits to those not owning property in the vicinity must be considered, and this benefit may constitute a large part of the whole.[49,50,51] Also, care must be taken not to include value changes due to other causes.[52]

It appears from the above discussion that economic approaches can be used to evaluate water-based recreation, but not most other environmental benefits. Other approaches described below may have broader application.

Public Opinion and Attitudes

Gilbert White feels that there has been a profound shift in the assessment of environmental values over the last few years, but he has no full explanation for it.[53] He points out that value systems differ greatly from group to group, partly due to cultural influences, and there may be a considerable difference between expressed attitude and actual behavior. For these reasons it is relevant to investigate the views of the public.

Interviews with different groups of users of parks have produced different rankings of motivations with the only similarity in motive cited as being that of escapism.[54] The principal conclusion to be gathered is that these users of the outdoors had divergent views as to the advantages to be gained, and found great difficulty in defining such views. If these views of outdoor living and water were to be amplified by investigating views of nonusers, an even more complex picture would undoubtedly be produced.

Index Approaches for Measuring Environmental Values

This sytem uses a system of index numbers of indicators, which are supposed to represent either magnitude of environmental impacts, social value of impacts of a given magnitude, or both. The most impressive index method of quantifying impacts was originated by Luna Leopold, et al., of the U.S. Geological Survey. The impetus for developing this detailed matrix relating "the actions which cause environmental impacts" with "the existing environmental conditions that might be affected" was the 1969 Environmental Policy Act.[55] In addition to an accounting of monetary benefits and costs, the act directed all agencies of the federal government to identify and develop methods and procedures to ensure that previously unquantified environmental amenities and values be given appropriate attention in decision making.

The method permits two values to be determined for each impact, each on a scale from one to ten. The first measures the degree to which the environmental condition is impacted, a ten reflecting total destruction. The second measures the degree of importance of the impacted environmental condition, a ten reflecting maximum importance.[56] At least this system forces planners to be aware of what they are doing to many "insignificant" environmental conditions. However, "the system is more of an inventory or cataloging system than an overall evaluation system ... and the freedom of selecting relative weights ... opens the door for the evaluator's bias to enter the evaluation, thereby leading to nonobjective evaluations."[57]

A variation on the Leopold procedure was developed by the Bureau of Reclamation in connection with analysis of the western U.S. Water Plan. This method is similar to the Leopold approach, however, "it provides no means for overall aggregated environmental evaluations." It concentrates on three main areas: quantity, quality, and human influence. "No attempt is made to discriminate relative importance of parameters in the three environmental dimensions, suggesting that this is left up to the evaluator's judgment."[5][8]

Indices are likely to be biased in some way in that the specified index numbers are developed by an individual or a group of specialists whereas a random sample of individuals would theoretically be less bias. We should not accept any single index or indicator, even the often-cited species diveristy, as fully determining the value of a biological community. However, indices may still be valuable in planning, since we must be prepared to accept some loss of precision for increased ability to communicate.

Governmental Agency Shadow Pricing

Governmental agency shadow pricing may be defined as estimates of money value usually made by economists to apply to items for which no market prices exist. In practice, when government agencies evaluate recreation they have usually followed the method outlined in Senate Document 97,[59] which specified two ranges of values to be used for recreation in project evaluations. General outdoor recreation, including swimming, boating and picnicking, was valued at $.50 to $1.50 per visitor day, while specialized outdoor recreation, including hunting, was valued at $2.00 to $6.00 per day. There are two weaknesses in this approach. In the first place, the valuation is subjective, although it was decided upon after prolonged discussions involving several years of time and hundreds of concerned persons. Probably more important, these unit benefits were applied to projections of future recreation demand, but the projections were usually made on the assumption of free entry by the public. However, if a fee of several dollars were charged, fewer people would use the facilities. Therefore, even if unit values were correct, aggregate benefits might be considerably overstated. In principle, this method is equivalent to the economist's concept of shadow pricing.

Interview Methods and Public Perceptions

This approach is used with the aim of avoiding the bias inherent in the opinion of experts by attempting to obtain directly the opinions of society through responses of a sample that represents as closely as possible the aggregate opinions of society. Each individual is asked the maximum price he would pay in order to avoid being deprived of a given amenity, or to obtain it if it is not already available. This technique is based on the assumption that the interviewed individual attempts to maximize his utility through rational allocations of his time and money.[60] The interview must be carefully constructed and conducted in order to avoid prejudicing its objectivity by the manner in which questions are expressed. If the individual expects to receive the goods in question in spite of his response, he may understate his value for it, especially if he expects to be charged for the value he indicates. If, on the other hand, the individual wants the benefit in question but does not expect to have to pay for it, as is often the case for government facilities, he may overstate his estimate of what it is worth to him.[61]

Methodology used and results obtained from interviews vary widely. Residents' opinions of water quality in streams differ sharply and nonsystematically from those of water quality experts. That is, the water quality as perceived by the public does not necessarily correspond to actuality. Nonspecialists relate to environmental aspects of water quality in terms of the following categories listed in order of perceived importance: (1) wildlife support capacity, (2) recreation opportunity, and (3) aesthetic aspects, composed of industrial wastes, cleanness and odor of water debris and algae.[62]

Other Measures of Environmental Quality

Although most of our top analysts have agreed that it is theoretically impossible to evaluate all aspects of environmental quality, many others, particularly some systems analysts and engineers, have no hesitation about undertaking such analyses. As already shown, there are acceptable methods of at least approximately estimating the economic value of outdoor recreation. Although aesthetics is often cited as a field beyond the scope of economic evaluation, in practice aesthetic qualities of landscape are evaluated in real estate transactions, art objects are sold commercially, and market prices for an original Picasso or Van Gogh are determinable. When an economist speaks of shadow prices he is in effect asserting the evaluability of nonmarket goods. It is possible that there are intangibles that cannot be evaluated in money terms at all, but this hypothesis has not been demonstrated definitively to be correct. Even the preservation of the last thirty whooping cranes might be given a money value, possibly in the range of one to ten million dollars, depending upon public opinion at the time the decision was made. Through the courts, methods have even been devised that in effect place an approximate value upon human life in terms of money.

These considerations merely conceptually illustrate the possibility of evaluation. Theoreticians who develop such approaches are not likely to be particularly well equipped to determine such values themselves, other than for the items evaluated by market prices. It should be noted that shadow prices are inevitably determined subjectively, that judgments regarding environmental quality may vary widely from person to person, and that the interview technique is never very accurate and can be misleading, depending not only upon the choice of a sample but also upon the techniques employed and the state of information of the public.

The systems analysts who are searching for general solutions for evaluating environmental quality are mainly offering schematic approaches filled out with unsubstantiated estimates. The methodology for making the difficult value judgments remains to be developed and will require the expertise of disciplines other than economics; the same observation applies to system engineering. Also, ecological analysis of environmental impacts is generally weak and lacking in data, and such analysis usually cannot adequately predict the changes in ecosystems likely to result from human intervention.[63] Research to remedy these deficiencies would be far more useful now than further elaboration of systems analysis approaches.

Economic Theories of the Counterculture

For a while assertions were made that pollution is an inherent consequence of our capitalist industrial development, so that a clean environment could only be obtained by a basic change in the political structure of our society. In fact, however, the origins of industrial pollution are clearly explained by the prevalence of economic externalities in the three-sided relationship between the polluting industry, its source of income, and the environmental interests of the public. According to Marshal I. Goldman of the Russian Research Center at Harvard University, "In our concern with pollution in the United States, we are sometimes led to believe that pollution is an inevitable by-product of capitalism. A study of the Soviet Union shows, however, that environmental disruption, in all the forms we are so familiar with here, also exists there. . . . The USSR finds itself abusing the environment in the same way, and to the same extent, that we abuse it."[64] Apparently the Soviet managers have given primary attention to economic production, just as we have in the United States, with a resultant failure to preserve the environmental interests of the public. Environmental damage has also been extensive in China for a long time.[65] The theory that pollution is an inevitable consequence of the capitalist nature of our society, which can only be overcome by changing the basis of that society, does not appear to be supported by analysis of western political economy or by the evidence available. However, such views are very convenient for those whose basic political views are anticapitalist, who are delighted to further an environmental theory which lends support to those views.

Aesthetics and Environmental Value

It is extremely difficult to evaluate the aesthetic element in man's appreciation of the landscape, because of difficulties of definition and also of individual aesthetic appreciation. To some water resource analysts *aesthetic* is simply a word used to characterize all noneconomic aspects of environmental value. However, there are many other social and cultural factors related to appreciation of nature and the outdoors, such as ecological values, companionship, escape from the pressures of society, fishing and hunting, and the simple opportunity for repose. Aesthetic values are more properly defined as those aspects characterizing man's perception and appreciation of the landscape itself, especially his visual perception, rather than its appropriateness for serving some human activity or cultural interest. Aesthetic values of the environment may be defined as those that are seen, heard, smelled, or felt, that are considered beautiful and pleasing.[66]

However, when we turn to general theories of aesthetics we find little that appears relevant to the natural environment. For example, Osborne[67] explicitly defines beauty in aesthetics as meaning the excellence of a work of art, as contrasted to the generally accepted meaning of beauty, which conveys emotive approval. Theories of aesthetics have been largely concerned with the long-standing question of whether beauty (or aesthetic merit) can be considered to have any objective quality or whether "beauty is in the eye of the beholder." Most analysts seem to feel that tastes vary so widely that there is no such thing as an objective aesthetic judgment. However, if this is so, what is generally considered bad taste may be considered equally as important and valid as what is considered good taste. This conclusion is embodied in the famous maxim *de gustibus non est disputandum* ("there is no disputing about tastes") but few writers on aesthetics are really satisfied with this viewpoint. Some insist that artistic experiences have an intrinsic value in themselves, and that art is its own justification.[68] Osborne[69] feels that some objective aesthetic reality exists, although imperfectly known.

Santayana[70] considers that "an aesthetic bias is native to sense, being indeed nothing but its need and potency." He feels that harmony in the natural rhythms and oppositions of life is developed by the aesthetic faculty. However, in character with the description of him as "the Last Puritan,"[71] Santayana feels that beauty must be restrained: "The fancies of effeminate poets in violating science are false to the highest art; and the products of sheer confusion, instigated by the love of beauty, turn out to be hideous. . . . To keep beauty in its place is to make all things beautiful."[72]

Although there were some opinions in the past that nature and art were similar and closely related, a modern trend among artists has been to shock, to rebel, and to express social movements. As Henry Moore said in "Unit One," "Beauty in the later Greek or Renaissance sense is not the aim in my sculpture."[73] Insofar as artists express their general views and ideologies through their works, they obviously represent something entirely different from natural beauty.

There is another aesthetic viewpoint that must be mentioned, and that is the love of architectural grandeur and luxury. In an extreme form these aesthetic manifestations were highly antisocial, as in the extravagantly magnificent gardens surrounding Versailles and other European palaces, and the twenty-five miles of uninterrupted view from the windows of the Vanderbilt mansion near Asheville, N.C., said to have been purchased in its entirety in order to safeguard it for the pleasure of future generations of Vanderbilts. Our ancestors were culturally conditioned to be impressed and overawed by huge aristocratic residences surrounded by spacious woods, parks and gardens that remain. The popularity of those converted to public use indicates that their physical characteristics still have meaning for the world, in spite of their historical origins in exploitation and inequity.

Despite the lack of a consensus of expert opinion as to what constitutes aesthetic value, in a general sense, various analysts and planners have made attempts to categorize the aesthetically valuable aspects of water-related landscapes. R.B. Litton, Jr., and associates prepared a comprehensive report for the National Water Commission of the aesthetic value which water contributes to the landscape.[74] They proposed a visual classification system, a set of descriptive categories, based upon three interrelated basic elements that together constitute a linked visual resource of water, land form, and vegetation. Visual components and relationships are classified. Three basic aesthetic criteria—unity, variety and vividness—are to be used as "generic means of identifying aesthetic quality." Three kinds of units are considered: the landscape unit, the setting unit, and the waterscape unit. Associated man-made manipulations can be enhancing, compatible, or degrading. The state of mind of the observer is considered important, including his intentions in visiting the area and his prior background. Human impacts are modifiers that may be in opposition to or in sympathy with the natural conditions and visual composition of the unit. This is a very complex approach. The relative dominance or subordinance of water is expressed in the setting. Isolation of the waterscape from its setting must be assessed as a lowering of quality, whereas movement is probably the most exciting and vivid quality associated with water.

More recently, an interview method using color slides produced the following conclusions as to natural beauty.[75]

1. A scene that includes a view of running water is usually preferred over one that includes still water or no water at all
2. The stark beauty of a desert, lava flow or a winter pasture is not perceived by most people
3. Some types of visual pollution (i.e., misfit billboards) are not recognized as such by some groups of people
4. Familiar scenes are not considered particularly beautiful even though they may be so to outsiders
5. Occupation and life style seem to have more effect on an individual's concept of natural beauty than age or sex

6. People usually agree on what is very beautiful or very ugly in a scene but disagree on the in-between

The above is only illustrative of approaches that are being made to this subject. The NAR report (North Atlantic Region interagency comprehensive plan, 1972) had an extensive classification of the aesthetic value of landscapes.

Nature modified by man can still be beautiful and attractive. Although some painters, such as Corot, glorify the natural landscape, there are many others whose paintings find beauty in village scenes, canals, boats and recreation spots. Homes and gardens on the Thames River (above the zone of pollution) and the well-planned river fronts in San Antonio are aesthetically delightful. Of course, many water-based recreation facilities provide fresh air, water and space for human activities without any particular beauty or, the beauty is submerged by overcrowding, but there are always advantages to be obtained by good planning.

One of the attractive features of the Rutgers University campus is a pond of about one acre in a large parklike open space, well landscaped with lawn and trees. This is certainly not a natural environment, as the pond was formed by an impoundment, its most apparent fish life is hundreds of small goldfish, and its other permanent inhabitants are a few varicolored ducks, crossbred between domestic ducks and wild mallards. In the winter and spring, dozens of wild mallards fly in and stay for months on end. Small boys come to fish for goldfish, students play and lounge around the pond, and dogs and strollers come by for diversion. There are a few chipmunks and a careful rabbit or two in the patch of woods alongside. It is a modified environment, but it produces a closeup view of wild mallards in mating season, which is worth quite a lot in the rapidly urbanizing region of northeast New Jersey. We may not know how to adequately define beauty or the environmental value of modified landscapes, but past examples, particularly in Europe, show in practice how it can be accomplished.

All aesthetic evaluations are necessarily subjective to a certain extent. Some may find beauty in a swamp, while others find it depressing and even ominous. Nonetheless the aesthetic aspects are important, and we must find better ways of evaluating the beauty of both natural and altered environments. Some criteria for doing this have been devised.[76,77,78]

Water Quality and the Environment

Other than taking the shoreland by development, water pollution represents the main impact of man's activities upon the water environment. Although our national water quality control program is now getting into high gear with an enormous construction program, the improvement in quality of our rivers is somewhat dubious, as indicated in Chapter 1.

Pollution manifests itself in unsightly conditions, odors, adverse effects upon biota, and, at times, menaces to public health. The aesthetic value of water and its surrounding geography is based primarily upon its appearance. At a minimum, this means water bodies should be free from obnoxious floating or suspended substances, particularly solids from domestic sewage, and objectionable colors, such as those resulting from some industrial waste discharges.[79] However, various analysts such as Harris[80] have shown that it is quite difficult to establish objective standards for color and odor. Table 2-1 shows great variability in the perceptions of water quality by different individuals. Figure 2-1 shows the variation of mean public acceptability, as related to changes in odor and turbidity. These aesthetic qualities of water are subjective and difficult to measure, but they are important to the public, even where the water is satisfactory from a health viewpoint.

Commonly accepted standards of water quality cited by the National Water Commission Staff[81] are as follows:

1. Nitrate over .9 mg/1   high nutrient levels cause significant
2. Total phosphate of .1 mg/1   problems
3. Suspended solids over 80 mg/l   cause significant problems
4. Total solids over 25 mg/1   cause moderate problems
5. Total coliforms over 1000/100 ml   unacceptable
6. Fecal coliforms over 200/100 ml   unacceptable

The by-product of power production thermal pollution, is also capable of

Table 2-1
Perceptions of Water Quality[a]

	Respondent Sample: Users of			
	Bottled Water (N = 20)	Filtered Water (N = 20)	Unfiltered Water (N = 20)	Total (N = 60)
Acceptable				
1	–	2	2	4
2	–	2	3	5
3	1	2	3	6
4	2	3	5	10
5	11	6	2	19
Not Acceptable				
6	2	4	3	9
7	1	–	2	3
8	3	1	–	4

[a]For the combination: turbidity 5, color 15, and odor 3

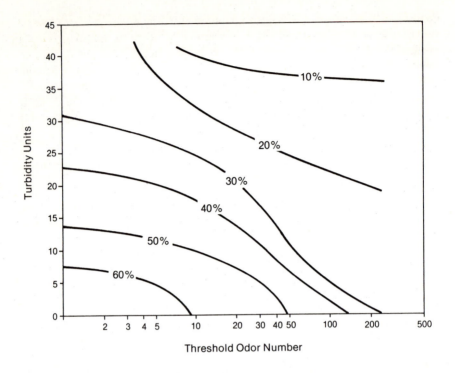

Source: Harris, D.H. "Assessment of Turbidity, Color, and Odor in Water," Santa Barbara, California: Anacapa Sciences, Ltd., for Office of Water Resources Research, 1972.

Figure 2-1. Acceptability of Odor and Turbidity Values

causing damage. Nuclear plants produce half again as much waste heat from the same power as do the traditional fossil fuel plants. Cooling water drawn from rivers, lakes, and estuaries has been returned to the waterway at temperatures as high as 115 degrees. The heat of water and uptake of dissolved oxygen are inversely related, which indicates another source of harm to fish. The toxicity of many inorganic wastes increases at higher temperatures and safe levels of toxins become critical as temperature increases. To complicate the already complex problem even more, sometimes heat is detrimental to certain kinds of fish. The states have regulated thermal discharges to prevent environmental damage.

It is clear that pollution is very harmful to fish life. According to the Federal Water Quality Administration (EPA's predecessor organization), over forty-one million fish were killed in 1969, more than in 1966, 1967 and 1968 combined. Apparently twenty-six million fish died in 1969 as a result of years of dumping by food-processing companies located at one lake in Florida.[82]

However, knowledge of toxic effects on aquatic life is very limited due to the complexities of research techniques involved. Also, evaluation of water quality is complicated by the nature of the receiving water. Its temperature, the levels of acidity and alkalinity, and the quantity of dissolved oxygen present constitute other variables which may synergistically add to the adverse effects.

There is an overwhelming spectrum of new chemicals that eventually ends up in water and a continual reuse of water as it travels downstream, resulting in a buildup of refractory materials that can contribute to degradation of environmental quality. U.S. Geological Survey monitoring stations feed into EPA's STORET (Storage and Retrieval System) information with respect to thirty-one parameters, of which the following eighteen are measured at over 1000 stations: temperature, specific conductance, turbidity, color, odor, pH (field), pH (lab), total dissolved solids, chloride, nutrients (nitrogen), nutrients (phosphorous), common ions, hardness, radio chemical, DO, BOD, coliforms, and sediment (suspended). These data are useful, but they are entirely insufficient to detect all of the pollutants which may be causing damage.

The environmental effects of pollution depend very largely on the ecosystems in the receiving waters and the uses to which the water is put. Broadly speaking, there are seven principal uses of water: (1) life support of fish and other natural species, (2) water supply, (3) contact sports such as swimming, (4) noncontact recreation, (5) navigation and disposal of wastes, (6) irrigation, and (7) cooling water. Some of the most important water quality requirements for these purposes, such as dissolved oxygen concentration, are well understood, and others, such as coliform count, have been defined clearly even though knowledge is very limited. However, there is a vast amount of ignorance about the effects of various chemicals alone and in combination, under particular circumstances, upon various aquatic species or upon various uses of the water.

A *New York Times* writer has recently emphasized the link between water quality and health hazards.[83] Outbreaks of disease from drinking water have been occurring at a fairly regular rate of more than one per month nationwide, although only a few get national attention. In the period 1960-1970, there were 130 officially recorded waterborne disease episodes in the United States resulting in 46,000 illnesses and 20 deaths, with the potential for such problems increasing. For example, residents of Miami Beach, Florida were jolted by word from county officials that they should boil their drinking water because contamination had been found in the supply system.

Known waterborne diseases include: (1) gastrointestinal disorders and infectious hepatitis, (2) eye, ear, nose and throat infections, and (3) skin infections. Sophisticated instrumentation and epidemiological techniques are producing indications that previously unsuspected diseases may be waterborne. A recent study[84] indicates that pathogenic strains of a free-living amoeba cause a type of meningoencephalitis, from which fatalities have been reported in six states in the United States and in several other countries. Moreover, there is concern that viral infections, particularly infectious hepatitis, may be contracted by bathing in sewage-polluted water. Human enteroviruses include the polioviruses, certain coxsackieviruses, and the echoviruses. Most of them produce disease only in a small proportion of the individuals infected. Diseases caused by virus include a variety of common respiratory ailments. Infectious hepatitis is the only fecally excreted virus for which definite epidemiological evidence of waterborne transmission has been found, but undiagnosed, low-level viral infections may infect other persons and thus initiate disease not readily attributable to water transmission. Viruses occur in very low densities compared to indicator organisms (coliforms), but they have been shown to persist for long periods and may resist chlorination somewhat better than coliforms.[85]

It would be desirable to provide a more rational health standard for water, to replace the present arbitrarily designated total coliform and fecal coliform counts, but considerably more research will be required before this can be done. The coliform standards for waters used for contact sports are quite varied. Total coliform standards have been adopted by thirty-five states and territories, ranging from 5000/100 m to 50/100 ml. The majority specify 1000/100 ml. Fecal coliform standards have been adopted by twenty-five states and territories, from 1000/100 ml to 70/100 ml. The majority specify 200/100 ml. A total of forty-eight states and territories have general prohibitions against any substances in water in concentrations or combinations harmful to human, animal, plant, or aquatic life.[86]

Other than as regards public health, impacts of pollution on the environment are usually visualized in terms of fish kills, and a valuable ecosystem is usually defined in terms of well-known fish, animals and birds. However, a river or lake ecosystem is much more complex than this, consisting of the following major classes of biota, united into an interrelated ecosystem:

1. Viruses, bacteria, algae, plankton
2. Water plants
3. Microinvertebrates
4. Insects, mollusks
5. Fish, reptiles, water fowl, and wild animals

Although the general public is well acquainted with selected fur-bearing animals such as rabbits, almost all fish, and birds, especially ducks, few people realize the complex interrelationships within a natural ecosystem. Ecologists have been hard put to explain in terms intelligible to the public just what value there is in a natural ecosystem, or even its consistency. One of the most basic elements is the use of the sun's energy to produce organic material through photosynthesis; this basic productivity is the originator of all energy flow, passing up the food chain by predation usually to larger or more complex forms of life, but passing also with decay of unutilized food or dead organisms to microorganisms or other scavengers. However, this type of analysis is limited in usefulness to planners, since some high energy-producing conditions (eutrophic) are not considered valuable.

A satisfactory ecosystem is said to be marked by long term stability, but this may include wide fluctuations or cycles of phases, for example, cycles of predator prey relationships. Rabbits or mice might build up a large population, followed by an increase of owls, which would consume the rabbits. Many owls would then die from starvation, and the cycle would then recommence.

Ecologists feel that a desirable ecosystem is usually marked by a large number of species and is characterized as having diversity. This argument seems better suited to inland waters than to forests. Most people have high regard for the extensive Western forests of sequoias, redwoods, Douglas firs, and other conifers, in each case occurring in stands almost exclusive of other large trees. In waterways, desirable environments with wide diversity of species are usually contrasted with undesirable environments having only a few dominant species, such as the Tubifex sewage worms that are sometimes packed by the thousands in the bottom. As an exception, the author has seen beautiful mountain lakes in the Rocky Mountains that seemed to have no other sizeable fish but trout. However, in more normal climatic conditions, it is certainly true that waters having a wide diversity of species seem to evidence more of the characteristics that we consider desirable. Where diversity is limited by pollution, either the less desirable species occur or those that occur are contaminated. The handsome, lively blue crab of clean bay waters is an attractive catch and very good eating, but the same species taken from polluted waters is stained, slimy and unsafe to eat.

A Canadian study was made to try to determine what aspects of the environment, not directly related to man's use, gave it its value.[87] The authors could not find a good answer, and ended only with reaffirming the classic doctrine that the degree to which the ecosystem is organized and the value of its natural life to mankind are positively correlated.

Discussion

The General Environmental Aim

Mankind in general is interested in nature, especially in combinations of wood, water, and grassy open space, in flowers, birds, wild animals, and even lower species such as frogs and caterpillars. That the combinations be aesthetic and unusual adds to the attraction. The modern environmentalist movement with its conceptually novel approaches has added vitality and color, but the essence of man's essential environmental interest is perhaps as well described by the nineteenth century concept of "Nature," as by any language of today.

Whether or not man's interest in natural environments is fully genetic or is partly cultured may be argued, but is not really very important. There is no assurance that present genetic characteristics are the best for any species, as, for example, the instinct which inclines a dog to chase automobiles. It is of course conceivable that by a process of natural selection and cultural change, a new generation of urban children could ultimately be bred who would prefer an urban setting and would not be attracted to natural environments. Various authors have hypothesized that the adaptability of the human race is so great that environmental conditions initially considered intolerable may become accepted and thereafter barely noticed;[88] however, as Dubos has stated, if this happened much of value might disappear from our civilization at the same time. Moreover, our value system is anthropomorphically centered, and the human race in question is that which exists today. However we may choose to explain it, there is a basic need of mankind, and particularly of children, to have frequent access to natural environments, either primeval or altered. Even plastic Christmas trees have symbolic value.

Some persons feel that religious feelings and ethical concepts are of key importance in understanding environmental relationships, but this is certainly not the case for children and probably not for the generality of mankind. The attempt of ecologist spokesmen to postulate environmental changes as basically immoral, so that an ethical mandate against encroachment can be postulated has not been successful.

The preservationist's view that endangered species and unique systems of vegetation are important to protect, regardless of their relationships to man's activities, is obviously cultural in nature. Except for true believers in special forms of polytheism, who must be excessively rare in current day society, it is hard to postulate any ethical objective not related to either man or God. As a practical matter, the preservationist's approach has much to recommend it, since it avoids the otherwise onerous task of explaining exactly why the whooping crane or the elephant seal is important to society. However, in the interest of clarity of thought, it seems better to keep in mind that natural species are important because of their interest to mankind, and it adds no ethical sanction to recall that various Indian tribes considered animals to have souls.

The idea of beauty of the environment has only within recent years obtained influential political backing.[89] Aesthetic concepts of environmental value undoubtedly have a broader appeal to adults than do religious views, although the aesthetic sense is probably not highly developed in children. It would be more plausible to postulate that the aesthetic views of adults grow from basic understanding and attraction for natural beauties. Aesthetics should have an important place in our planning, but we must be careful to avoid elitist views. The landscaping ideals of the French chateaux, and of some of the American castles of later vintage may have great aesthetic value; but less refined and less grandiose approaches are more appropriate for public planning in our present society. Aesthetic values are largely cultural, and may change with time. Environmental aesthetic values cannot be evaluated directly.

Economic Approaches and Public Action

Although man's attitude toward "Nature" is a basic part of his genetic cultural heritage, it must be noted that when men are hard pressed for survival, environmental interests are given secondary importance when compared to the primary urges of self-preservation. Our remote ancestors, being cold and hungry, gave primary importance to food and shelter and to safety from enemies, and in similar circumstances, so would we. During World War II, the British cut down magnificent beech groves to make tank parks and potato farms. The undeveloped nations cannot be counted on to establish wild game preserves on a large scale when their populations are short of agricultural land. When hard economic decisions have to be made, environmental quality has always had to give way to economic necessity, but in the United States, at least, we now have opportunities to create an environmentally improved country without impacts so severe as to impair our ability to maintain our population in the near future. We may conclude that although the desire for a good environment is less basic than other instinctual drives, such as that of self-preservation, it can in principle be compared to and measured against other social needs, and against marginal economic utility, in a society in which the minimum needs of existence are generally obtainable.

Conclusions

The majority of environmental benefits are very poorly understood by philosophers, planners or decision makers, and certainly we are far from having either the technology or the factual data with which they may be fully evaluated. For the time being, the theoretical approaches, which assume a total evaluation of environmental aspects, are many years away from full application and may be regarded as having dubious value. Inconvenient though it may be, there is no rationally based procedure currently available to evaluate all of the environmental benefits numerically. Recreational values can only be estimated and index approaches can give only approximate indications of other environmental values.

3 Effects of Pollution on Receiving Waters

It is an error to conceive of the natural waters of our country as ever having been entirely pure. Although many springs and mountain streams were, and still are, of very high quality, containing only harmless minerals and dissolved gases which improve the taste, other waters are naturally charged with excessive amounts of substances in such high concentrations as to be considered polluted. Many examples can be cited: the Mississippi and Missouri Rivers were heavily loaded with silt when Lewis and Clark first saw them; the Arkansas River at Keystone carries chlorides amounting to over 10,000 tons daily, in concentrations of 600 mg/l, which are largely of natural origin; the lower Arkansas, before its impoundments were built, carried 100 million tons of sediment a year; and the Kissimee and St. Johns Rivers in Florida have unsatisfactory dissolved oxygen due to natural causes. However, ecosystems have developed and become adjusted to these environmental conditions, and when new conditions are artificially imposed on them, especially added pollution, many species find it impossible to adapt themselves and are eliminated. The result is a changed ecosystem, which generally turns out to be less desirable and less useful than the original. The term "pollution" should only be used when referring to undesirable quantities of substances in water. Of course, the question of how much is undesirable remains to be decided.

Pollutants include many substances which are difficult for biota to absorb and cycle, being too abundant or too toxic to be assimilated. Thus, they may accumulate in lakes, estuaries, marshlands or in sediments of flowing streams. Some of these substances, such as heavy metals, radioactive materials, and persistent organics, may remain to pollute the environment for many years to come. Pollutants now accumulated in sediments cannot be assumed to represent long-term stable concentrations. It is much more likely that in most estuaries and lakes the accumulations of persistent pollutants represent mostly the residue from recent years, and will continue to increase indefinitely unless a very large degree of control is exercised. Although the quicker-acting nonaccumulative toxics, the effects of which are rapidly dissipated, may cause the most spectacular adverse effects on the environment, such as major fish kills, the persistent buildups are of great long term importance, and are much less well known.

A succession of unexpected events has warned us that modern industrial progress poses more environmental dangers than we can foresee in any detail.[1,2] First we had unfortunate effects of DDT, then mercury, and then polychlori-

47

nated biphenyls (PCBs). Now we have found that two common building materials, asbestos and polyvinyl chloride, may be cancer producing, and restrictions on their release into the general environment appear necessary. In the United Kingdom, the wool industry uses polyglycols, which have now been found to be toxic even in minute quantities[3] and a recent serious episode in the United States involved the manufacture of kepone pesticide in Hopewell, Virginia. It has become apparent that there are serious environmental hazards associated with a wide variety of chemical substances, and not all of them are known. Continued attention to toxics will be necessary in any system of water pollution control.

Dr. W.S. Becker in a 1973 editorial in *Science,*[4] expressed concern for the lack of knowledge regarding impacts of pollution in receiving waters. He spoke of the unpredictable response of algal plant life to the presence of phosphorous and other nutrients which we are attempting to limit, indicating that in many cases, unduly high sediment loads, subtle toxicities, or turbulence from flow may be the controlling influences rather than the nutrients. He states that the fate of toxic metals, radioactive substances, organic wastes and intestinal bacteria is even less completely known. Means of monitoring and analyzing the processes involved are yet to be developed. Pending basic research on these aspects, he considers that ecosystem computer modelling is largely without a factual basis, and its predictive power is questionable.

Substances need not be toxic in order to constitute pollution. Water contaminants can be classified as:

1. Physical constituents (suspended solids)
2. Biological constituents
3. Chemical constituents (dissolved solids and gases)
4. Color, odor and taste
5. Toxic substances
6. Nutrients
7. Thermal excess
8. Acidity (or alkalinity)
9. Radioactive contamination

Outside of radioactive contamination, which is relatively rare and is monitored and controlled entirely separately, the other forms of pollution named are widespread and are generally characteristic of unprotected streams in metropolitan areas.

In water to be used for drinking, the EPA, under the Safe Drinking Water Act, has now outlined interim limits for turbidity, bacteria, some pesticides and inorganic chemicals, such as lead, mercury and arsenic, and has established combined standards for organic chemicals. Some of the maximum contaminant levels are shown in Table 3-1. Behind these apparently precise levels there

Table 3-1
Maximum Contaminant Levels

Inorganic Chemicals		Organic Chemicals 0.7	
		Pesticides	
Arsenic	0.05	Chlorinated hydrocarbons	
Barium	1.00	Chlordane	0.003
Cadmium	0.01	Endrin	0.0002
Chromium	0.05	Heptachlor	0.0001
Fluoride	1.4-2.4	Heptachlor epoxide	0.0001
Lead	0.05	Lindane	0.004
Mercury	0.002	Methoxychlor	0.1
Nitrate (as N)	10.0	Toxaphene	0.005
Selenium	0.01	Chlorophenoxyls	
Silver	0.05	2,4-D	0.1
		2,4,5-T	0.01

actually is a great deal of uncertainty as to the exact effect of various contaminants.

Sediment

Sediment is one of the most significant of pollutants. It is popularly referred to simply as "silt," but clays, sands and organic substances are important components. The United States has carried on a national program of soil conservation for over forty years, with the primary aim being to reduce erosion of soil. Despite these efforts, sediment remains one of our fundamental water pollution problems, and measures of the 1972 Act have not mounted much of an attack upon it; they have only affected the portion that comes through waste treatment systems.

A sediment load in water creates turbidity which cuts down light penetration and thus limits production of algae. Since algae are among the important food sources in a stream or lake, reduction of the fixation of carbon by photosynthesis greatly curtails the productivity of the food chain. On the other hand, where nutrients are high, excessive algal productivity often exists, in the form of algal blooms. The curtailing of algal growth by turbidity is one reason why undesirable algal blooms do not develop in river systems to the extent that they do in lakes.

Sediment from urban runoff occurs mostly during storms in concentrations running to hundreds of parts per million, which usually subsides to negligible quantities during normal periods. Sediment on the bottom often clogs up spawning grounds for valuable fish, and covers over and destroys the small bottom creatures (benthal fauna) that live there. Some suspended sediment and dredged material is not inert at all, but contains heavy metals, petroleum, or other chemicals. It is particularly harmful to those organisms that filter their food out of the water or sediments.

It has been shown by Leopold and others that cutting down a forest may increase the sediment load entering the river system eight times, and an increase of cropland from 10 to 50 percent may increase the sediment four times.

The straightening of a meandering stream channel greatly increases its capacity to carry heavy sediments. The shortened river has an increased slope, which increases the sediment-carrying capacity. (Carrying capacity for bed load increases approximately as the third power of the product of depth times slope.) Moreover, the elimination of bends by straightening reduces the hydraulic losses associated with changes in direction. Accordingly, in a straightened channel, the scouring power of a given flow of water is greatly increased. If the stream was originally clogged with sediment it may now carry the load more efficiently; but if it was previously in equilibrium it may now tend to degrade its bed and erode its banks to a greater extent than it did previously.

The "fate" of sediment is defined as the changes in the substance after it enters the stream. Fine particles may be carried indefinitely by a stream's currents as suspended sediment, but the heavier sedimentary particles soon descend to the bed of a stream, to make up what is called the bed load. These particles either remain in place awaiting the next high water or move intermittently along the stream bottom, producing an unstable substrate which is unsuitable for most bottom dwellers. In small streams draining urban and urbanizing areas, the suddenly increased rush and greater total volume of storm runoff, due to impermeable paving and roofs, greatly accentuate the sediment carrying capacity of the stream. If the stream originally was in approximate long-term equilibrium between the erosive force of the waters and resistance of its bed and banks, the new conditions tend towards rapid erosion, which once started, moves upward to the smallest headwaters until more resistant material is encountered. Such processes often lead to remedial programs of channel lining, which destroy the environmental value of a natural stream.

The huge quantities of suspended sediments in a stream are reduced in quantity in passing downstream. On the average, a small river basin of one thousand square miles in area only discharges from its lower end from six to twenty percent of the sediments poured into it from its banks and minor tributaries. See Figure 3-1. What happens to the other eighty to ninety percent? No adequate explanation appears to be available. In some cases, the basin may be in a state of geologic disequilibrium, such that sediments moving down from the headwaters during the last fifty years or so are accumulating within the channels of the streams, and only much later, decades from now, will sediment be carried forth from the lower end in full force. However, the volumes of

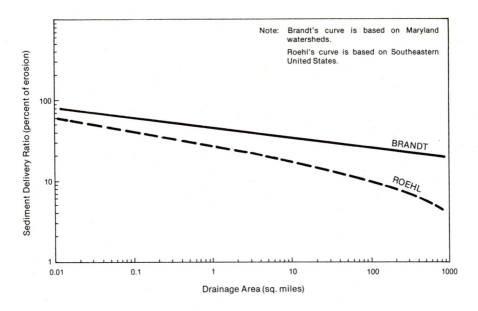

Source: From Chen, C-N, in "Urban Runoff Quantity and Quality," American Society of Civil Engineers 1975, p. 162. Reprinted with permission.

Figure 3-1. Effect of Drainage Basin Size on Sediment Delivery Ratio

sediment are too great relative to channel capacity, and there are too many basins involved, for this theory to be a general explanation of the reduction in sediment mass. It must be that as sediments move down a stream the particles are ground together physically, or chemically disaggregated in some other fashion, much more effectively than we can now explain. This is a fascinating subject for research, but for our present purposes, it is sufficient to know that the relationship does exist.

Dissolved Oxygen Deficiencies

Dissolved oxygen concentration (DO) is a very important factor in water pollution, and one to which a great deal of attention has been given. It was realized very early that dissolved oxygen in water is extremely important to fish life. This effect can be stated very simply: At levels of DO below 2 mg/l most fish cannot survive for any appreciable period of time; below 4 mg/l a healthy diversified fish population cannot be maintained; below 5 mg/l a healthy trout fishery cannot be maintained. These relationships vary somewhat with other characteristics of the water, and vary considerably with the duration and frequency of the oxygen deficiency, and with its timing with respect to the life cycle of the fish; but these limits are good general guides. Dissolved oxygen is the nearest thing we have to a general indicator of water quality in streams.

A stream has a given biological community to support. When the flow is reduced, uptake of oxygen is correspondingly reduced, and may become insufficient for the biota. It has been roughly estimated that a stream flow of ten percent the average flow will sustain short-term survival of aquatic life. Less than that corresponds to severe degradation.[5]

Biochemical Oxygen Demand

This is the main process affecting the dissolved oxygen level of streams. It represents the combined demand for oxygen by chemical and biological oxidation, by chemical compounds and by bacteria, respectively, in the water over a period of time, under specified temperatures. When not specified, the period of time is understood to be five days. That is, a BOD of 10 mg/l means that that particular water will use up 10 mg of oxygen per liter of water, during the next five days, if held at a temperature of $20°C$. At any other temperature, demand would be higher or lower, in accordance with definite relationships. In polluted waters, the demand is usually exerted by the action of many different bacterial species. However, this complexity need not concern us, because for most ordinary conditions bacteria are so numerous, and multiply so rapidly, that conversion of the organic material can be assumed to be limited only by the amount of material available, and not by the numbers of bacteria. The process follows a first order relationship, which is expressed in simple mathematical terms:

$$\frac{dL}{dt} = -KL \qquad (3.1)$$

Where L is the remaining ultimate BOD in mg/l, and K is a coefficient depending mainly upon the temperature and upon the character of the polluting material. As a first rough approximation, it can be estimated that an organic pollution loading in summer will usually decrease by about one-third every twenty-four hours.

This BOD relationship, derived many years ago, and used for almost all water quality modelling ever since, must be modified by consideration of four other aspects: (a) validity of the first order relationship, (b) benthal oxygen demand, (c) photosynthesis, and (d) nitrification.

Validity of the First Order Relationship

The first order BOD relationship was derived primarily for use with raw sewage, which is very readily biodegradable; it does not apply as well to industrial wastes, treatment plant effluents or to polluted water in streams, which have substantial proportions of materials quite resistant to bacterial action. In these cases the situation is apt to be somewhat like that shown in Figure 3-2. The total demand from organic material in these cases (according to the first order mathematical relationship) should amount to 1.5 times the five-day BOD, if the test is run long enough. However, it usually amounts to something like 2.0 times the five-day BOD in twenty days. This means that the total oxygen demand exerted within twenty days (which is a time of residence exceeded in many estuaries) is one-third greater than indicated by the usually assumed relationship.

Moreover, other circumstances can reinforce this discrepancy. If toxic substances exist, or the wastewaters tested have been chlorinated, the long term oxygen demand may be 10 or more times the five-day BOD. Moreover, oxygen demand from ammonia is usually not indicated by this test.

From time to time dissatisfaction is expressed with BOD as a measure of

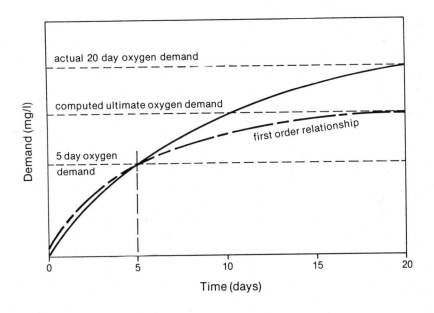

Figure 3-2. Biochemical Oxygen Demand

organic content of water, and attempts are made to substitute some more readily monitored characteristic such as COD (Chemical Oxygen Demand), or TOC (Total Organic Carbon). The main difficulty is that five days are required to determine BOD in a laboratory, whereas a number of other characteristics can be determined instantaneously in continuous monitoring stations. The measurement of BOD values below 2.0 mg/l is questionable, particularly where algal growth is abundant. Algal contents of lightly polluted water provide an oxygen demand consisting of algal respiration rather than bacterial action on organic matter.[6] However, no satisfactory substitute for BOD has yet been found, although considerable modifications in its use appear likely in the near future.

Benthal Oxygen Demand (Oxygen Demand from the Bottom)

Bottom muds in slowly moving waters are generally anaerobic. The thin surface layers of these muds are high in bacterial activities, which exert a considerable oxygen demand on the water above. This is called benthal oxygen demand, and it must be considered in any complete examination of the oxygen regimen of the river.

Even a body of water which has an appreciable content of dissolved oxygen in its waters may have anaerobic muds. Under anaerobic conditions in either waters or muds, most of the normal bacteria which utilize organic material are replaced by anaerobic bacteria, the life processes of which are entirely different. The utilization of organic material is much slower, noxious odors (including hydrogen sulfide) are created, and nitrogen gas may also be produced. These gas emanations are caused by processes which remove oxygen from nitrates and sulfates, especially as the dissolved oxygen approaches zero. When the organic content of sediment deposits is high, the sediments may remain anaerobic for years.

In one case, Hunter found bottom sediments which burned freely when set alight; almost one-third of their total dry weight was petroleum.[7] These sediments were found in a small tributary of Arthur Kill (west of Staten Island, N.Y.), which drains perhaps the greatest petroleum refining complex on earth. Of course, such extreme conditions are rare, but it is usual for bottom muds in slowly moving waters to be generally anaerobic, with thin surface layers of the muds high in bacterial activity, exerting a considerable benthal oxygen demand on the waters above.

Photosynthesis

Within the waters of an aerobic stream, the phenomenon known as photosynthesis introduces a daily cycle of changes in the oxygen level. Photosynthesis is the basic process of plant life, by which carbon dioxide and water, in the presence of sunlight and nutrients, are converted to organic substances, with the liberation of oxygen. This process is presented in more detail in the discussion of lake eutrophication, in this chapter. Here it is only necessary to note that the plant life in streams (mostly minute algae but also rooted aquatics) produces oxygen through photosynthesis in daylight hours and uses up oxygen during the night hours, a process known as "respiration." This process results in a cyclic diurnal fluctuation in DO concentration. In many streams of considerable turbidity the fluctuation amounts to only about 1 mg/l daily, but if nutrients are present in sufficient amount and the water has low turbidity (so that sunlight can penetrate), the diurnal fluctuation can be considerable. The usual net effect of the diurnal photosynthesis cycle is to add more oxygen during the day than is used up by respiration at night. However, at summer temperatures, when the maximum solubility of oxygen in water may be only 8 mg/l, large oxygen fluctuations may exceed the saturation level during daylight, and oxygen may be lost in the air. In one extreme case, the oxygen went to 22 mg/l in the daytime and descended to 2 mg/l at night. Obviously, in such a stream, conditions are quite unfavorable for fish, in spite of very high average DO concentrations. Since most water sampling is conducted during daylight hours, the presence of marked photosynthesis results in overestimating the concentrations of DO available, considering that fish must breathe throughout the night as well as during the day.

Nitrification

A fourth complication affecting the DO regimen comes from the existence of the nitrification process in polluted streams. Ammonia is a component of sewage and of much other pollution. It is quite toxic to fish in concentrations of 1 mg/l or above, and is also important as an oxygen demanding substance. A given quantity of ammonia demands about 3.8 times the same quantity of dissolved oxygen to convert it to nitrate. On the Upper Passaic River, concentrations up to 3 mg/l of ammonia have been found; this represents the same ultimate oxygen demand as a BOD of about 11.4. However, ammonia released in streams acts in what seems at first a rather erratic manner. Sometimes, practically the entire ammonia content will be oxidized to nitrate within a few hours of its release. More characteristically, in streams of moderate size, the flow will continue for days on end with no appreciable nitrification at all, and then the process will be carried on actively in the estuary. To understand these differences, it is necessary to understand something of the bacterial processes which cause them.

In laboratory tests the nitrification of ammonia usually is manifested only after six or seven days; therefore it does not show up in standard five-day BOD tests. (This is another reason for underestimation of oxygen demand by standard

BOD tests.) However, if the process is continued for twenty days, a double curve is indicated, as shown in Figure 3-3, demonstrating the extent of oxygen demand due to ammonia.

Nitrification of ammonia occurs through only two kinds of bacteria, *Nitrosomonas*, which convert ammonia into nitrite, and *Nitrobacter*, which change the nitrite into nitrates. Both types of bacteria are relatively slow growing. Also, they have a very strong tendency to grow on surfaces, rather than freely in the water. Therefore, in spite of otherwise favorable conditions, the numbers of the right kind of bacteria in the water are ordinarily insufficient to produce appreciable nitrification in streams. However, in shallow, rocky and weed-grown streams, which have a relatively high wetted surface area, or in a trickling filter, which is a component part of some waste treatment plants, the nitrifying bacteria accumulate in huge numbers and are able to process the ammonia to nitrates in a very short time. In estuaries, the water is deep, but travel of water masses is slow, and there are usually many suspended particles caused by contact between fresh and saline waters. While research is just beginning, it is probable that the nitrification bacteria cluster on these suspended particles, and that the slow travel time is sufficient to allow the process to be completed within the estuary.

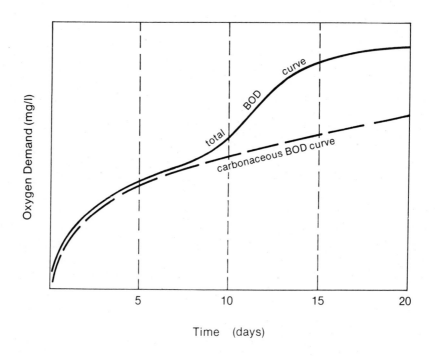

Figure 3-3. Carbonaceous and Total BOD

The nitrification process may have a bearing on undesirable eutrophication of lakes. The process provides one of the essential nutrients for algal growth, although the limiting nutrient in inland waters is usually phosphorous, or occasionally nitrates or silica. There is usually one nutrient in limited supply for algal growth, the others being present in sufficient quantity.

Nitrates in estuaries are especially important, because nitrates are usually the limiting nutrient to biological growth in estuarial waters, as contrasted to phosphorus, which is usually limiting in inland waters. In some estuarial waters, and in the sea, an increase in nitrate is usually desirable.

Modeling Oxygen Demand in Rivers

Returning to the matter of dissolved oxygen in streams, it is necessary to employ an analytical and predictive approach to establish the relationship between various organic loadings entering the stream (BOD) and the resulting dissolved oxygen concentration. This was first done many years ago, after noting that the natural aeration from the surface tended to counterbalance the oxygen demand created by any pollution loading. This natural aeration is effective proportionately to the extent of the oxygen deficiency, below a maximum or saturation value of dissolved oxygen. The combined relationship of pollution loading and natural aeration is called the Streeter-Phelps Equation, and forms of it have been used in modeling ever since its derivation; it is expressed mathematically as

$$-\frac{dL}{dt} = K_1 L - K_2 D. \tag{3.2}$$

Where L is the ultimate oxygen demand in mg/l, and D the oxygen deficit (saturation concentration of oxygen in water less actual concentration). K_1 and K_2 are coefficients. The equation is equally valid where L is given in pounds, representing loadings. If used to describe stream conditions downstream from a single large pollution source (of BOD = L), a characteristic oxygen "sag curve" is produced, as shown graphically on Figure 3-4. One of the curves indicates a moderate decline in DO and the other a decline to zero DO (anaerobic or septic conditions) followed in each case by a recovery of DO as the biochemical demand is expended, and the natural forces of aeration again become controlling.

The basic relationship for modeling dissolved oxygen must of course include all major factors. In an estuary, matters become extremely complicated, on account of tidal flows, possible stratification between fresh and salt water, etc., however, in nontidal waters, the assumption is usually made that a pollutant is dispersed uniformly across a stream, that flow is steady, and that conditions are uniform from the surface of the water to the bottom. The benthal demand and the diurnal DO variation due to photosynthesis must be superimposed in order

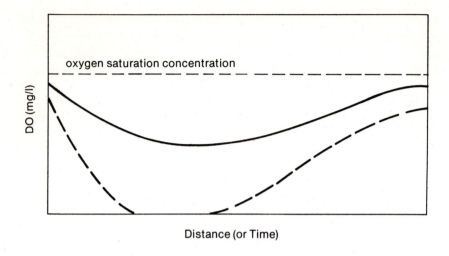

Figure 3-4. Oxygen Sag Curves

to obtain true minimum and maximum values. In such cases some form of the Streeter-Phelps equation can be formulated, tested and verified with pollution loading and other data (taken at periods of relatively uniform flow) as follows:

$$-\frac{dL}{dt} = K_1 L - K_2 D - \overline{(P-R)} + B \qquad (3.3)$$

where $\overline{P-R}$ represents the mean value of photosynthetic production minus respiration, and B is benthal demand.

By use of such a model, if all the pollution loads entering a stream are known, the effects of DO of future changes in these loads can be evaluated, including changes due to load growth and to various alternative waste treatment alternatives.

The modeling of dissolved oxygen is widely practiced. It is very useful to administrators, because it is sufficiently complicated a process to impress the courts in a judicial proceeding, when introduced by an expert witness. It is profitable to consulting firms, because it is not very difficult for trained personnel to carry out. Modeling is an essential and useful scientific approach, however, it is subject to abuses. When introduced with appropriate solemnity, it allows a mediocre professional to gloss over serious deficiencies in data, and even gross misapprehensions as to the processes involved, while gaining great respect from the majority of his clients and the public. The uncritical acceptance of the process has resulted in an overreliance upon low-flow DO modeling, and a neglect of other equally important forms of analysis, such as those involving effects of nutrients, sediment, heavy metals, and pollution from unrecorded sources and storm runoff.

Although it is generally accepted that the critical dissolved pollution conditions in rivers occur at times of low flow this is not always the case. For example, in the Kissimmee River, Florida, little pollution comes from point sources, and dissolved oxygen deficiencies occur mainly at high flow.

BOD from nonpoint sources and urban runoff increases more or less proportionately with the discharge. The capacity of a river to aerate itself to cope with the oxygen demand of pollutants varies with its surface area and with the water velocity and turbulence. Large quantities of heavy metals and sediments in streams are washed in during heavy rains; accordingly, the most important conditions governing these pollutants are those of storm runoff. Increases in river discharge correspondingly dilute any continuous discharge of pollutants. A reduction in pollution concentration may be caused by unpolluted flows from an undeveloped watershed, or discharges increased by reservoir releases. Conversely, reductions in flow may increase pollution effects. This is illustrated by the modeling of the Willamette River by the USGS. As shown by dotted lines in Figure 3-5, with the same BOD loading, a reduction of summer discharge from 5000 cfs to 3000 cfs would reduce the DO level from forty-three percent saturation (about 3.9 mg/l) to eight percent (about .7 mg/l).

Public Health

In this country a great deal of attention is given to coliform counts as indicators of potentially dangerous bacterial infection. However, in Great Britain, after prolonged study, mainly at beaches, no connection was found between bacterial counts and public health, and the test was abandoned for bathing waters. The British now swim in great numbers in a side channel of the Thames River at Oxford, and in the even more polluted Serpentine (a small lake) in Hyde Park in London. Although the murky waters doubtless would fail U.S. coliform count standards by large margins, there seem to have been no serious consequences. The Norwegians also have followed the British lead in this respect. On the other hand, in this country various illnesses and infections have been traced to polluted water, usually when ingested, but also from contact while bathing. Most authorities consider that the fecal coliform count is the more reliable indicator, but the total coliform count is still more widely used. Despite the uncertainties involved, there seems to be no move to abandon reliance upon coliform counts in the U.S.

Heavy Metals

Heavy metals were brought prominently to public attention a few years ago, mostly as a result of cases of mercury poisoning (Minamata disease), and the

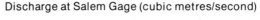

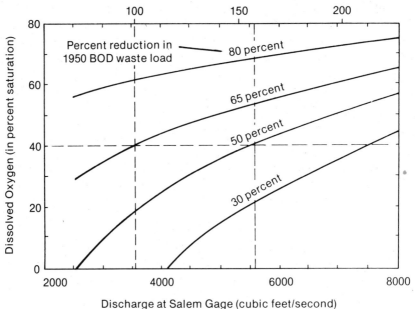

Source: Adapted from U.S. Geological Survey, Circular No. 4, 1975.

Figure 3-5. Relationship between Pollution Loading, DO, and Flow, Willamette River

discovery that certain fish, particularly swordfish, had mercury concentrated in their bodies to much greater levels than in the waters and the bodies of food chain organisms. However, other heavy metals occur much more extensively, and their importance is just beginning to be realized. The heavy metals occur largely in storm runoff from urban and industrial areas and are generally associated with sediments. Therefore, the material can drop out and occur as benthal deposits, even though concentrations in the water itself may be low. According to Dr. Ruth Patrick,[8] recent research indicates that heavy metals discharged in parts per billion may accumulate in many thousands of times the solution concentration in certain algae and invertebrates. This may be transferred up the food web in harmful amounts, and may inhibit growth of species of benthal fauna (i.e., bottom-dwelling species, such as snails, worms, and larvae of mayflies and other insects). In some cases benthal fauna develop the capability of absorbing and retaining high concentrations of heavy metals, and are then eaten by their predators. Such processes help to explain the general lack of desirable species of fish in the rivers of urban and industrial areas.[9] As another consequence of the association of heavy metals with sediments, the practice of filtering water samples prior to analysis may result in seriously underestimating the heavy metals content of the water.

Effects of Phenols and Chlorinated Hydrocarbons

Among the more toxic pollutants commonly found in streams are phenols and various chlorinated compounds. Phenols occur in petroleum, are created by various chemical manufacturing processes and, in very small concentrations, are even created by natural processes. They are relatively persistent as regards biological degradation. From studies of the EPA[10] about the problem of water-polluting spills of oil and other hazardous substances, a priority ranking of soluble hazardous substances was developed, and phenols were first on the list.

Phenolic compounds are toxic to fish and shellfish in concentrations of around 1 mg/l, or even much less.[11,12] There is also the distinct possibility that phenolic compounds may be absorbed by commercially used species such as oysters, whereupon their taste may be affected. A recent EPA sponsored study of the effect of water pollutants upon flavor of fish showed that of twenty-two organic compounds tested a chlorophenol had the greatest effect, having a threshold concentration of only 0.4 ppb (parts per billion).[13] The abatement of pollution in the upper Ohio River following closure of the steel industries on July 15, 1959, led to a marked improvement in water quality accompanied by a rapid increase in the variety and abundance of fishes.[14] During the shutdown, phenolic contents averaged 3 ppb, whereas the content was increased to 16 ppb after the mills resumed operation, with subsequent deterioration in water quality. Although phenol content was much less than usually specified toxicity levels, it appears that considerable harm was being done.

Chlorinated hydrocarbons may occur from petroleum sources in two ways: (a) direct discharge in wastewaters or effluents, and (b) formation from the chlorination of wastewaters or effluents. Chlorinated hydrocarbons that may occur from these two sources have been indicated by Russian sources to have "safe" levels of from 0.002 to 0.2 mg/l concentration, based upon combined studies of toxicity, clinical aspects and effects upon water quality.[15,16,17,18,19,20,21,22] A few studies have shown sensitivity of organisms to phenolic compounds in concentrations as low as 0.001 mg/l.[23,24,25,26] Obviously these substances are generally objectionable even in minute quantities.

Waste Treatment Effluents

The occurrence of objectionable chlorinated compounds in our rivers is largely the result of chlorination of sewage effluents, which is now generally required.

Potentially carcinogenic materials which have been discovered recently in the drinking waters of Cincinnati and New Orleans are believed by EPA spokesman to be originating in the chlorination process.[27,28] Municipal wastewaters contain hydrocarbons, and the chlorination process can produce chlorinated hydrocarbons.[29] Some of this class of chemicals are extremely toxic, as indicated above, but the extent to which chlorination actually produces specific toxic compounds from hydrocarbons remains to be determined. However, an extensive study of municipal wastewaters in the San Francisco Bay area indicated significant increases in toxicity following chlorination.[30,31]

A recent report summarizes effects of chlorinated effluents from 156 secondary sewage treatment plants in Virginia, Maryland and Pennsylvania.[32] Total chlorine appeared to be the best indicator of reduced fish diversity in receiving waters below the outfalls. Bioassay results show chlorinated bi-phenyls to have much greater effects on algal communities than on pure cultures of algae in the laboratory.[33] The toxicity of chlorinated effluents is a matter of immediate research concern, in view of the emphasis which has been given to chlorination of all wastewater effluents, and the tentatively indicated EPA policy that urban storm runoff also should be disinfected.

In stressing the toxic effects of phenols, heavy metals and chlorinated wastewaters, we must not overlook the probability that there are in wastewaters many other toxic substances as yet unknown. The effects of two combined sublethal stresses may be more than the sum of the two effects, if the stresses were applied separately, and some of the unknown toxics may react synergistically to increase the adverse effects of known toxics. Various attempts have been made to simplify, combine or average water quality parameters in order to obtain general indices of water quality. However, all such attempts are of dubious value. If fish are killed by one pollutant, it is immaterial that all other water quality parameters may have been satisfactory. The most critical parameter for a given purpose is that which measures the utility of water for that purpose.

Heat

Thermal limitations for discharges from large installation have been fairly strictly enforced. Of course, the disastrous effect of hot industrial wastewaters on small streams has been noticed by almost everyone and control of such waste-heat releases is essential. But there is some doubt as to the justification for the exact limitations now being imposed on large industrial and utility companies. The two most damaging fish kills that have occurred recently due to heat, in the writer's experience, have been due to the heat being suddenly withdrawn. In the case of the Oyster Creek nuclear power station, the Atomic Energy Commission required a suspension of operations in midwinter for safety reasons. Numbers of fish, which were attracted by the unseasonably warm water and had remained in the vicinity were unable to adjust so suddenly to the cold waters of the Atlantic seaboard and died of thermal shock. A similar case involved a winter shutdown of operations on an inland power station. The heat itself was obviously attractive to the fish.

Even the change of a few degrees in the mean temperature of a river might result in major ecosystem changes if the input were constant; this does not mean that there would be no fish. New species of fish similar to those found in abundance in other rivers further south could no doubt be introduced. However, environmentalists find such changes highly objectionable, as explained in Chapter 2.

Biologic Assimilation of Pollutants

As Ruth Patrick brought out, a water body has a certain assimilative capacity for pollutants.[34] Wastes of many kinds can be broken down by bacteria, fungi, and some of the invertebrates, and the resulting waste products can be utilized by algae and other forms of aquatic life. If these algae are of the species that have high grazing (herbivore) pressure, there will be an efficient transfer of energy through the food web, ending up in high productivity of fish desirable for sports and commercial fisheries. On the other hand, if the same amount of energy is converted into algae which have low herbivore pressure and thus accumulate in great numbers, not only will the energy transfer through the food web be greatly curtailed, but nuisance growths of algae will develop; the lake, river or estuary will become unsatisfactory for fishing as well as aesthetically unpleasing.

Recent research has shown that extremely small amounts of trace metals such as manganese, vanadium, nickel and selenium may greatly affect the whole structure and the efficiency of algal functioning at the base of the food web.[35,36] It is necessary to have these trace metals in the right concentrations in order to have algal species that are an effective food source. Pollution may suddenly kill organisms by interfering with vital enzymatic or nutritional pathways, or pollutants may accumulate in an organism and produce chronic effects on the creature itself, or on predators which feed on it.

The health and continued existence of organisms depends upon the aggregate of the stresses which they receive. It is not always possible to predict the combined effect of two or more stresses. Sometimes one stress will serve to neutralize another, as most obviously in the case of alkalinity and acidity. Most often the adverse effects of stresses reinforce each other. For example, Haskin found that oysters succumb to lower concentrations of petroleum pollution when they are poorly fed.[37] Sometimes two stresses will act synergistically, that is, the combined effect will be greater than the sum of the two effects considered separately. The rapidity of change is also an important factor, particularly as regards the impact of unusually low or high temperature.

Species tested by bioassay frequently show a much greater sensitivity to pollutants at certain stages of the life cycle, particularly egg and larval stages. For example, Haskin found that oyster spat (the free swimming larval stage of oysters) are killed by low concentrations of petroleum to a much greater extent than are adults; moreover, many of the surviving spat are unable to set (establish themselves in a fixed location as a metamorphosed infant oyster).[38]

In general, only a bare beginning has been made in evaluating the true effect of pollutants on the ecosystem. A great many laboratory experiments have been carried out, mostly to determine the short time lethality of a single pollutant on a single species. Research on chronic effects and life cycle studies and effects upon communities has been very limited. Information such as the high heavy metals counts in the bottom creatures, or benthal fauna, in Greenfield, Mass., discussed in the preceding chapter, indicate that the origins of pollution and the mechanisms of biological effects of pollution are both very imperfectly understood.

From time to time new toxic effects of pollution are discovered, indicating that considerable hazard may be lurking in the general ignorance which still encompasses the subject of environmental effects. For example, it has been discovered that polychlorinated biphenyls, which are waste products of some industrial processes, accumulate in the bodies of fish from rivers and lakes, even when the concentrations in the water are very low. Ingested by humans in sufficient quantities, they accumulate in the fatty tissues of the body and may result in a variety of toxic effects, including death.

Eutrophication of Lakes

Eutrophication of lakes has long been known to occur as a natural process, the symptom of geologic aging. In this process, natural lakes slowly accumulate organic material and nutrients and become more productive of plant life, ultimately perhaps filling up altogether to form a marsh or swamp. As the process gets well under way, with production of considerable plant life, the lakes are known as eutrophic. (Those lakes free of eutrophic conditions are characterized as oligotrophic, and in between states are called mesotrophic.) Figure 3-6 shows a eutrophic canal.

As shown in the preceding chapter, man's activities result in greatly accentuating the flows of silt and nutrients into lakes, and consequently accelerating the progress of eutrophication. Most of the small and medium sized lakes and impoundments in metropolitan areas of the United States are more or less eutrophic. This means at various times of the summer and fall, especially during the summer, the presence in the water of excessive numbers of microscopic plants called algae, gives the water a murky look of varying hues of brown, gray and green. However, many lakes in the less settled parts of the

Figure 3-6. Eutrophication in Water Supply Canal

country are in better condition, the land being less densely occupied. For example, of 2790 lakes classified in Wisconsin, eighty-seven percent of the total shoreline had a development density of fifteen or less dwellings per mile.[39]

It must be emphasized that "eutrophic" and "noneutrophic" are relative terms. There is no definite borderline. In the long run there is a spectrum of conditions, in which certain lakes have more rapid "aging" processes than others. However, there is a great variability of conditions from time to time in any given lake.

The most widely known lake eutrophication problem in the United States is that of Lake Erie, and it has been grossly misrepresented. As Ketchum notes, the frequent statement that Lake Erie is "dead" is biologically absurd, as it is the most productive of the Great Lakes (contains the most life).[40] However, this does not mean that its condition is a desirable one, since fish species produced in Lake Erie today are not those most valuable for man's use.

At times the algae in a lake may multiply very greatly and form mats or scums on the surface, called algal blooms. The algae then die, and the decomposition of their remains may exhaust the oxygen and result in fish kills.

There are many kinds of algae, and much remains unknown of their life cycles and the influences which control them. The class known as diatoms requires silica, and in lakes where they predominate, silica occasionally may be the limiting nutrient, the availability of which limits the extent of algal growth. However, such situations are uncommon. Varieties of green and blue-green algae are more commonly the dominant species in inland lakes, in which case phosphorus is usually the limiting nutrient.

The blue-green algae seem to form the most obnoxious algal blooms, sometimes causing bad odors, and even occasionally occurring in varieties poisonous to cattle which drink of waters heavy in algae. Blue-green algae are prey to certain viruses, which can rapidly annihilate all but a minute fraction of a huge population; however, there always seems to be a resistant algal strain present, which multiplies with incredible rapidity to replace the old one. The action of known virus species is probably only one of numerous factors which explain the great difference in algal growth situations under apparently similar conditions. As an additional complication, the blue-green algae often live in symbiotic relationship (that is, a mutually beneficial community existence) with certain bacteria, and themselves have some of the characteristics of bacteria.

Most lakes of the United States, outside of the extreme north and mountainous parts, are stratified in summer. This means that the deeper layers of the lake are cold and stagnant, and usually also totally lacking in oxygen. (Although some lakes, known as oligotrophic, may be stratified but have oxygen at all depths.) Of course no fish, insect larvae, or shellfish can survive under such conditions. However, the upper twenty feet or so of medium sized lakes are oxygenated (aerobic). These upper waters are agitated and stirred by winds, and inhabited by fish and other normal marine life including algae. See Figure 3-7 (Lake Stratification).

In many lakes, while the lower water masses have no oxygen, the upper oxygenated layers of water are too warm for trout. A few trout may survive the summer in the middle layers, or they die out entirely and must be restocked in the spring.

As processes of eutrophication continue in smaller lakes, marshes, and impoundments, the surface may become covered with floating plants such as the tiny milfoil, or further south the water hyacinth or water chestnut. Moreover, rooted aquatic plants often begin to sprout from the bottom, to form masses of vegetation even in waters four or five feet deep, so thick in some cases that it becomes laborious to move a canoe through them. In some communities the

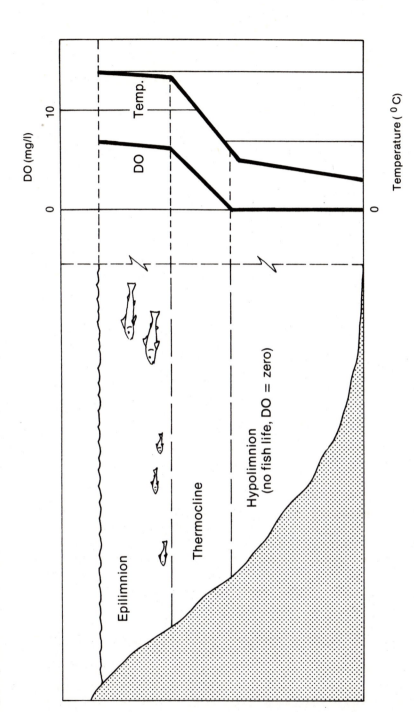

Figure 3-7. Lake Stratification

people become exasperated and pay for expensive dredging programs; Princeton University has dredged out Lake Carnegie, twice, twenty-five years apart, removing something like a million cubic yards of sediment each time. In past years, huge quantities of more or less toxic chemicals were dumped into lakes in attempts to kill the weed growth. It is now generally considered, however, that the best basic approach to avoiding eutrophic conditions in lakes is the control of the supply of nutrients, plus excavation of mud banks in shallow water.

Nutrients in Lakes

The pioneer in relating the eutrophication process to the nutrient budget was undoubtedly Sawyer, in 1947, who suggested threshold levels of inorganic phosphorus and inorganic nitrogen as limits on growth of algae. The nutrient levels are determined prior to the start of the summer algal growing season. Levels suggested were .01 mg/l P and .3 mg/l N. If both were exceeded, growth would be expected to be excessive. These limits were derived for lakes in Wisconsin. Reported by Uttomark.[41]

Now, thirty years later, it is generally agreed that inflow of nutrients into lakes should be controlled; but there is still considerable disagreement as to the results to be expected from given nutrient reductions, and even greater disagreement as to the processes involved. It is quite apparent that there are far greater complexities in this matter than was imagined even a few years ago. However, further consideration of Sawyer's approach by Vollenweider[42] and subsequently by Uttomark[43,44] has confirmed the conclusion that the overall total phosphorus and nitrate annual budget (aggregate inflow) generally is decisive as regards the extent of eutrophication. It should be noted that although dissolved orthophosphate is the form immediately available to the algae, in view of the many conditions under which one form transforms into the other, total phosphate represents the potential for growth.

The permissible loadings which were first worked out took no account of "residence time" or the length of time a given mass of water remains in the lake, which has a considerable effect. An impoundment on a large river may be refilled with water a number of times each year, and so it could accept a much greater loading of nutrients than a lake of similar size with only a small outlet, or, still worse, no outlet at all. To take two conspicuous examples, Lake Tahoe has a water replacement time of about 700 years, so that it is many times more susceptible to eutrophication than Grand Coulee Dam, which has the entire upper Columbia River flowing through it. The same relationship holds for smaller impoundments.

Acceptable phosphorus loadings for avoidance of eutrophic conditions in lakes, giving consideration to residence time, have been worked out by Vollenweider, as reported by Dillon.[45] The total phosphorus loading for the entire year is considered the determining factor, related to the depth and the flushing time of the lake. The permissible and "dangerous" loadings are as indicated on Figure 3-8. In this figure Z is the mean depth of the lake, in meters, and Tw the mean residence time in years of water in the lake. Z/Tw is therefore the total depth of water in meters which would be added to the lake during the year if none were discharged. This simple empirical relationship has been widely accepted for planning purposes. For example, in Ontario the phosphorus loading approach of Vollenweider and Dillon is being applied to obtain indices of algal potential (chlorophyll A concentration) and of turbidity (Secchi disc readings).[46] Some much more complex models of the phosphorus cycle are being developed to predict the relationships in more detail. In applying these newer models, however, various investigators have found them to be somewhat unreliable, since it is often found that the parameters and relationships used do not encompass the totality of the influences upon algal growth.[47]

A survey of technical literature sponsored by EPA[48] indicates that if the

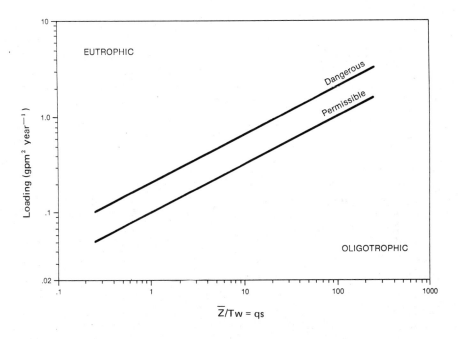

Source: Adapted from Vollenweider, as reported by Dillon, P.J., "The Application of the Phosphorous Loading Concept to Eutrophication Research," Burlington, Ont.: Canada Centre for Inland Waters, undated.

Figure 3-8. Acceptable Phosphorous Loadings

Nitrogen/Phosphorus ratio in water exceeds fifteen, phosphorus is apt to be the limiting nutrient. However, if either nutrient remains dissolved in water in significant quantities during an algal bloom, that nutrient is not apt to be limiting. The limiting nutrient can be determined more definitely in any case by bioassays performed on bottles of lake water, with and without the addition of more of the nutrient being considered ("spiking"). If addition of more nutrient than is already in the water results in a markedly higher algal growth, that nutrient is limiting.

Bottom sediments in lakes may accumulate large quantities of nutrients, which are released to the water. The processes through which this occurs are very poorly understood. It is thought that the anaerobic condition of the depths of the lake favors such a release, and iron compounds undoubtedly play a part in the action. However, it seems probable that this is not only a chemical process. Bacterial action appears to be implicated in some way, as chemical inferences from various sets of data are difficult to reconcile. There is a symbiotic (community) relationship between blue-green algae and bacteria which considerably complicates matters. Nutrients in water are continually assimilated by organisms and thereafter released either as excreta, as food to predators, or as remains. Despite the very considerable number of competent scientists at work on the matter, it is more than ever true that there is only limited understanding of the exact processes involved in lake eutrophication. The one conclusion we can be quite sure of is that the continued uncontrolled loading of nutrients into lakes will sooner or later render them eutrophic. In metropolitan areas, most lakes are already relatively eutrophic except some of the very large deep ones, which have not yet had time to be changed by pollution loadings.

Control Methods

Some of the crude chemical controls of the past have already been mentioned, such as treatment of eutrophic lakes by alum, or copper sulphate. Harvesting of weeds and algae has at times been attempted; most notably in herculean labors by the Corps of Engineers to clear water hyacinths out of southern navigation waters. Manatees and a special type of carp will devour aquatic weeds, and these species have been employed, the former with some local success, in Florida. The use of selected virus to eliminate blue-green algae has been suggested, but research has shown that the genetic flexibility of the obnoxious blue-greens can outflank the viral attack in a very short time by developing resistant strains. This occurs so rapidly that this viral approach does not appear promising. Deepening a lake can solve the problem of rooted aquatics, at least until the lake silts up again, and it should help somewhat with the algal problem.

In most cases the only permanent way of controlling eutrophication is by limiting the input of at least one nutrient, preferably the nutrient which is already limiting, usually phosphorus compounds. Cutting off direct pollution was very successful in controlling algal blooms in Lake Washington, although normally this may be difficult to accomplish. Waste treatment plants can treat wastewaters to remove phosphorus, but at considerable expense; improperly functioning septic tanks near any water body usually contribute to its eutrophication and may have to be rebuilt or replaced with sanitary sewers; and the principal means of reducing phosphorus input from cropland are by control of soil erosion and limiting usage of fertilizers. After some forty years of national effort by the Soil Conservation Service, there is still inadequate control of sedimentation from cropland, particularly in cases where land is rented. Feedlots and dairies require special attention.

Rather than relying exclusively on the nutrient loading guidelines discussed above, it is desirable to have better methods to predict eutrophic conditions in future situations. Many consulting engineers can offer mathematical models which purport to predict eutrophication, usually based upon phytoplankton activity, but such models should be viewed with great reserve, in view of the many uncertainties remaining in this matter.

Bubbling air of oxygen through anaerobic layers of stratified lakes has been suggested as a means of reducing the solubilization of phosphates contained in bottom sediments. A number of researchers, including the author, have been involved in experiments of this nature, but all such research so far has been quite inconclusive on account of unforeseen complexities in the situations encountered.

Turbidity in lakes is considered aesthetically undesirable. However, the control of suspended sediment to improve the clarity and appearance of the water might well accentuate obnoxious algal blooms. This could occur by penetration of light to deeper depths, thus accentuating the process of photosynthesis.

Estuaries ·

Estuaries are submerged river valleys, subject to tidal action, where the river occupies a larger valley than it could have scoured out under present geological circumstances. The most economically important estuaries in the United States are probably those on the East Coast, and three of them, the Hudson, the Delaware and the Chesapeake-Potomac are regions of really serious water quality problems.

Superficially an estuary resembles a river, but it is much more complex, both physically and biologically. Upper portions of an estuary are apt to be fresh water, the flow of which is reversed periodically by the tide. If the lower portions are fairly wide, they are quite saline. In some estuaries, such as the Delaware, the mixing of fresh and salt water at any point is fairly complete; in others, the salt water penetrates many miles upstream beneath the fresh water, in a long tongue, called the salt water wedge. In the latter case, the salt and fresh water masses only mix slowly. In very wide estuaries, the water masses are under the influence of the Coriolis force, due to the circulation of the earth, which causes them to rotate slowly, in addition to movements caused by wind and tides.

The fresh water flowing in the river must eventually find its way to the sea, but, because of the large cross section of the estuary, the net time of passage through the lower estuary is very long. Any given water mass may move back and forth for months under tidal influence before passing out of the mouth and mixing irreversibly with the ocean. As mentioned above, this provides ample time for nitrification to take place if free ammonia exists.

The heavier water-borne sediments carried down into the estuary, such as sand and gravel, are apt to be deposited quickly, and thereafter only move under influence of storms or unusual floods. In the wide expanses of an estuary, heavy wave action occurs, which may move the bottom sediments intermittently. The salt water mixing with the fresh has a tendency to flocculate suspended clays, causing them to settle. If there is much water navigation, the powerful screws of the ships and barges create turbulences, which keeps sediment largely suspended in the main channels; but the finer sediments settle out in any quiet waters, creating mud banks.

Marshes are apt to adjoin the estuary, due to the geologic land settlement which formed the estuary itself. Near the coast, other marshes are created by the same processes of sand movement as those which create coastal islands, with their beaches and bays. The salt marshes are usually covered with water at high tide and are overgrown mainly with tall marsh grasses. They are interpenetrated by tidal channels, which bring in water with the flood of the tide and carry it out on the ebb. Many local and migratory birds use the marshes as feeding, breeding or resting grounds. Marshes are highly productive of bacterial and vegetable matter, absorbing nutrients and organic matter from incoming tides and, at appropriate times of the year, flushing out bits of dead vegetation and decay products. Eminent ecologists are convinced that the marshes are valuable nurseries for important varieties of fish, and for the food chain species upon which they feed; but the marsh ecology is so complex and varied that only fragmentary information seems to be available to confirm this conclusion.

From early colonial times, man has developed the marshes for his own purposes; large areas were diked to keep out salt water, in order to facilitate the growing of hay. More recently, all kinds of land developers have filled in the marshes to build various structures, often combining the filling with excavation of channels, along which rows of homes could be built, each with its own boat dock or pier. The stagnant channels which result are convenient for boats, but they do not fulfill the ecological functions of marshes.

Many marshes have been drained to eliminate breeding grounds for mosquitoes and stinging marsh flies, or the marshes have been sprayed with insecticides, regardless of the damage to the abundant life which they contain.

The Corps of Engineers has diked many marshes and filled them perhaps fifteen feet deep with dredge soil. This process ultimately creates economically usable land, but it eliminates the marsh.

The public view of marshes is a mixed one. Probably the majority of the public considers them aesthetically unattractive and useless, but, to many persons they have great beauty and attraction. As regards pollution control planning, marshes have to be considered from two entirely different viewpoints. In the first, they are a natural resource, however dimly comprehended, which must be protected from indiscriminate development. Secondly, marshes serve a useful function in absorbing nutrients and pollutants of various sorts, and converting them into biologically useable and harmless forms. The extent to which this can be done without harming the marsh is not known, and is a useful subject for research

The biological life especially adapted to brackish waters is confined to relatively few species. There are in the sea enormous numbers of salt water species, a few of which can tolerate brackish water, like the oyster and the blue crab. Other salt water species, like the salmon and the shad, ascend far into fresh water streams to spawn. Similarly, some of the fresh water species can descend into brackish water. However, the total number of species which prosper in the brackish water of lower estuaries is only a fraction of the number which are either at home in fresh water, or at home in salt water.

Nutrients in Estuaries and the Sea

The eutrophication problem which is so prevalent in lakes is much less marked in estuaries, perhaps because so much of the vegetable growth is apt to occur in marshes. Undesirable algal growths may occur in open waters, but are less characteristic than in lakes. The limiting nutrient to algal growth in estuaries is apt to be nitrates rather than phosphates, and it is usually not clear in a given case whether or not a further increase of nutrients would be desirable. As Bostwick Ketchum has indicated,[49] the open ocean is extremely short of nutrients, to the extent that with ninety percent of the total ocean area, the open ocean produces less than one percent of the total fish catch. Coastal zone areas average about seventy-five times as much of a fish catch as equivalent areas of open ocean, and limited up-welling areas, where deep nutrient-laden ocean waters rise to the surface, which are still richer in nutrients, are much more productive still. However, as Ketchum warns, pollution can bring excessive nutrients which can replace the normal phytoplankton population with obnoxious species (cultural eutrophication).

Modeling of Estuarine Pollution

The upper portions of estuaries act hydraulically much like large rivers, except that they flow intermittently first in one direction and then in the next. They are kept mixed vertically from top to bottom by turbulence, just as rivers are, and also, with a greater lag time, they are mixed from side to side, and have considerable lag in their lateral mixing. A given inflow of pollution is apt to flow down for several miles at least before becoming mixed throughout the estuary cross section. On the assumption that this mixing process prevails, mathematical modeling of the estuary can be conducted as though all of the pollution in the estuary were concentrated along a straight line, regardless of its actual dispersion throughout the estuary cross section.

There is still one very large difference between rivers and tidal estuaries—the tidal movement creates added dispersion. At the end of a tidal cycle a given water mass has moved a long distance downstream and a somewhat shorter distance upstream. It is, however, much more thoroughly mixed with adjacent water masses upstream and downstream than it would be in a simple movement downstream of the same distance corresponding to the net travel in an estuary. The adjustments to take care of this dispersion in a mathematical model are considerable, but can generally be managed at not too great an expanse.

The biggest handicap to estuary modeling is that the measurement of pollution loadings in tidal waters is extremely difficult. The net pollution loading is the difference between total pollution movements up and downstream respectively, during the tidal cycle. Since these two relatively large totals can each be measured only approximately, the true difference between them may well be masked by the errors inherent in such estimates. Indirect approaches for measuring pollution loading are usually used, based upon rather complex hydrodynamics approaches, which are beyond the scope of this book.

In view of the large size of major estuaries relative to any given pollution source, and the difficulties of measurement discussed above, it is very difficult to estimate pollution loading entering the estuary from tributaries and unrecorded sources. Major point sources such as waste treatment plant effluents must be sampled directly. For estimates of urban runoff and nonpoint sources, it is necessary to sample a number of storm sewers and minor tributaries, using methods outlined in the next chapter. Results may then be extended to other areas, by comparisons of population and land use. Where important combined sewers exist, their loads must be measured individually, unless reliable data can be obtained as to engineering aspects of individual sewers. Lacking such detailed information, there is no basis for any assumption that two combined sewer systems will have similar pollution loadings.

Sediments in Estuaries

The bottom sediments of an estuary are important to its ecosystem. Not only waste treatment plants and major discharges but also urban and industrial runoff and unrecorded sources sweep pollutants out into our estuaries, much of which settles to the bottom as heavy sediments. The soluble and very fine particulate components may remain suspended in the estuary water for weeks or even months, and eventually pass out to sea. Much suspended material, including matter newly floculated by salt water, settles to the bottom as a sort of ooze, which is readily disturbed by currents or passing water craft. In the long run, bottom sediments may have a tendency to move down the estuary, enroute to the sea. However silts and other soft materials, once compacted, are not readily moved again. One of the principal activities disturbing such deposits is the periodic dredging of channels and anchorage areas for navigation purposes. The disposition of the spoil from dredging should be carefully controlled to prevent the fine material flowing back with the water into the waterways again, since it is often loaded with pollutants which are best left in the spoil disposal area.

Lower Estuary Mathematical Modeling

The mathematical modeling of the lower portions of a major estuary is a task which may cost millions of dollars. In the broad reaches of the lower estuary, the circulation of water is influenced largely by winds, tides, density, currents and the Coriolis force, and two-dimension two-layer, or even three-dimensional modeling may be necessary for accurate results, with a tremendous field effort required to obtain information necessary for calibration and verification. Such a model is being prepared to analyze organic pollution in New York Harbor and adjacent coastal waters (the New York Bight), but ordinarily water pollution control planning cannot support such efforts.

4

Urban Runoff and Nonpoint Sources

In the traditional approach to water pollution control, the main objective was to clean up raw wastewater and gross industrial wastes. However, now that secondary treatment of wastes is becoming widespread, the influence of other pollution sources must be taken into account. Official water pollution control programs have been directed entirely towards control of recorded wastes; however, the polluting nature of nonpoint sources and especially urban runoff is now being seriously studied. Within recent years many analysts have asserted the highly polluted nature of urban storm drainage.

Background

The earliest serious treatment of nonpoint sources and urban runoff, with real data, appears to be that of Weibel,[1] published in 1964. Later Whipple, in 1970, published findings that in three New Jersey river basins, the known point sources of pollution accounted for less than one-third the observed pollution.[2] Bryan in 1971 published studies of pollution from urban runoff, the findings of which are quite comparable to findings of Weibel and Whipple.[3] New York State officials had meanwhile become concerned about the problem. In 1972, the Water Pollution Control Act included the requirement that areawide pollution control planning should include consideration of both point and nonpoint sources of pollution. However, the EPA did not yet include nonpoint sources among the subjects for which they would consider funding research, and it was not until May 1973 that the EPA accepted unrecorded pollution as a potential research project. Efforts of the EPA to make up lost ground are described later in this chapter. In studies of potential treatment methods the EPA has funded good work, but in investigating the origins of nonpoint sources of pollution, its environmental effects, and its incorporation into planning processes, it has been slow and relatively ineffective. Only recently has the EPA emphasized the importance of storm water runoff and nonpoint source pollution. At the start of the 208 Planning Studies (funded by EPA), most of the available data and technology had to be taken from published and unpublished research reports funded by the relatively small research programs of the Office of Water Research and Technology, Department of the Interior, and other non-EPA sources. However, the EPA funded at least one excellent study of urban runoff pollution described later in this chapter.

77

Unrecorded wastes may enter the stream carried by seepage, overland flow, gullies and minor tributaries, but often also by pipes, such as storm sewers, combined sewer overflows, and minor industrial outlets. Small amounts of background pollution may be attributed to natural soils, rocks, and vegetation—under certain geological conditions large amounts of pollution, particularly saline, come from natural sources—and somewhat larger amounts result from raising of conventional row crops on level ground and from single family suburban housing. However, by far the greater quantities of most pollutants originate in urban and urbanizing areas, with many streets and parking areas, multiple residences, and especially commercial and industrial facilities of all kinds. Figure 4-1 shows an area with pollution from many such sources.

An analysis of state water quality strategies in 1975 found that the current national perception of nonpoint pollution sources included the following, in order of priority.

1. Feed lot, dairies
2. Sediment and erosion
3. Mining
4. Urban runoff—bacteria
5. Pesticides and agricultural runoff
6. Oil fields
6A. Nutrients—agriculture related
7. Land fills

Other sources mentioned were forestry practices, septic tanks, chloride intrusion and dredging. It seems that there was no concern expressed about any pollution from urban runoff except bacterial. This illustrates how slow has been the national recognition of this problem.

The origins of nonpoint pollution and urban runoff are diverse. Important nonpoint pollution sources including mining wastes; wastes from active mines are usually regulated, but acid mine drainage from abandoned mines and salt flows from abandoned and unplugged oil wells constitute a continuing problem. Irrigation return flows and drainage from animal feed lots are important waste sources, which have often been regulated. Commercial fertilizer applications, pesticide applications, refuse dumps and soil cultivation are sources of nutrients, sediment, and various other pollutants. Rainfall may be acid, or may contain objectionable quantities of sulphate, or mercury, or cadmium, etc. Automobiles leave hydrocarbons from oil and lead from exhaust gases in the air and on or near highways. Pesticides and nutrients from fertilizers wash off of cropland, or seep into groundwater. Groundwater may also become polluted by saline intrusion, sewage and industrial waste leakage and by leachate from solid waste

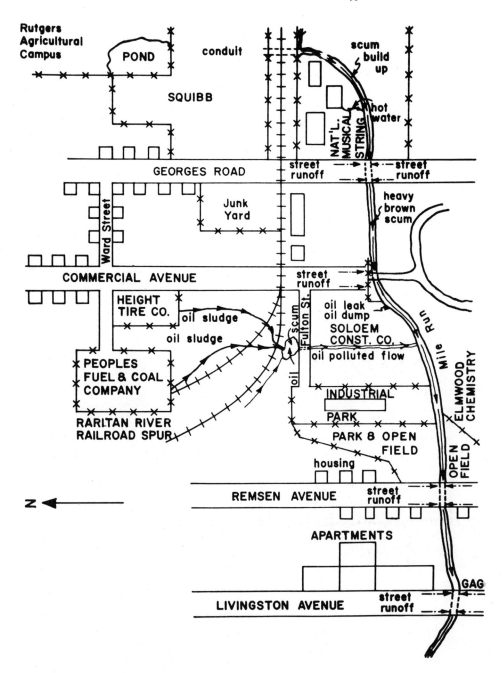

Figure 4-1. Potential Sources of Urban Runoff Pollution, Upper Mile Run

disposal. For example, recent studies have indicated that houses having septic tanks located within 300 feet of a street or swale may be associated with a deterioration of water quality.[4] Nine million tons of salts are spread on highways annually for deicing, and must eventually find its way to surface or ground waters. Additional salt comes from natural sources and from irrigated agriculture. As an extreme example, the average annual salt output attributed to irrigated agriculture in the Grant Valley of the Colorado River Basin is estimated at 460,000 tons annually, or over eight tons annually per irrigated area.[5] Recreational and industrial watercraft originate various kinds of wastes, of which oily bilges and garbage and sewage from cabin cruisers and houseboats are the best known. Bacterial and viral contamination come not only from human wastes but also from animals, both domestic and wild. A three-pound duck has been estimated to have a fecal condition output greater than the sewage from five adult men, and a feedlot raising 10,000 head of cattle is said to produce the same sewage disposal problem as a city of 165,000 population.[6] In fact, animal wastes of livestock in the United States are estimated to total about two billion tons annually, which is equivalent to ten times that produced by humans.[7]

Erosion contributes vast loads of sediment, largely from construction, lumbering haul roads and agricultural activities. Removing trees from a mountainside or forest vastly increases erosion, and the sediment is washed into the river.

The greatest of the pollution sources in the United States, other than treatment plant effluents, are those from urban areas, such as urban and industrial storm drainage and combined sewer overflows. Forrest Neil states that in the Chicago Metropolitan Area, more than forty percent of the pollution reaching the river system comes from combined sewers, in addition to pollution from urban runoff.[8] The Corps of Engineers has found that urban runoff entering heavily urbanized Peachtree Creek in Atlanta is sufficient to cause violation of stream quality standards quite frequently.[9] The Delaware River Basin Commission considers that, although the amount of pollution entering from stormwater overflow and tributaries of the Delaware River is not accurately known, the best estimate is that it was originally nineteen percent and will increase to forty-two percent of the total after present treatment plans are carried out.[10] An excellent study made by Rickert of the USGS on the Willamette River found that forty-six percent of the dry weather BOD loading of the river originated from nonpoint sources, even though much of the drainage area is forested and undeveloped.[11] An EPA spokesman indicated that, in cities where secondary treatment of wastes is provided, storm generated discharges account for from forty percent to eighty percent of oxygen demanding materials.[12] Some states including New Jersey, Pennsylvania, and North Carolina have officially recognized the importance of the problem, at least at policy levels. The latest annual report of the EPA shows great interest in nonpoint source problems.[13] Researchers in the field are happy to note the growing recognition of the problem, and hope that more technology can be developed, and that we can move from the stage of acknowledgement of the problem to the stage of doing something about it, at both federal and state levels. A recently initiated series of plans under Section 208 provide a means to do this, if the opportunity is taken.

Urban storm runoff is ordinarily very high in suspended sediments, with mean concentrations in the hundred of milligrams per liter, usually rising dramatically with peak flows. Coliform counts are also extremely high and very variable, often much in excess of concentrations considered safe for water contact activities. These facts are widely recognized. What is only recently becoming known is that very large quantities of BOD, heavy metals, nutrients, hydrocarbons and probably other pollutants are found in urban runoff. The methodology of determining concentrations and loadings of these substances, predicting future occurrence, and preventing or controlling resulting adverse effects is covered in the balance of the chapter.

Methodology

Most data gathering and modeling of water quality has been done for low flow, steady state conditions, on sizeable streams, and is done on the assumption that the critical period for pollution is the period of lowest flow. This is not necessarily the case. For large rivers and tidal estuaries in developed regions, the lowest dissolved oxygen levels are apt to come when, with a prevailing low flow, urban runoff from a summer storm produces large amounts of organic pollution. Also, low flow conditions are not determinative of lake eutrophication problems, which depend upon the total annual inflow and outflow of nutrients, and especially upon inflow of nutrients during flow conditions. Sediment and heavy metals from urban and industrial runoff also seem to be carried largely in storm flows, and industrial discharges may be seasonal in nature. Therefore, analysis must consider much more than a low flow condition.

The routine periodic sampling of small streams does not appear to give satisfactory values of the pollution loadings they carry. Theoretically such monitoring should include a proportional sampling of storm events, but it is unpleasant to be outside sampling during storms, so, unless close supervision is exercised, the important high flows are apt to be minimized in the record. Moreover, the flows from storm sewers and minor tributaries draining urban areas are particularly flashy, and the concentration of pollutants changes rapidly.

The loading may vary by two orders of magnitude or more in half an hour. The only way to be sure what is entering the stream is to employ very carefully designed automatic samplers, as most commercially available samplers are inadequate, or conduct hand sampling of the stream or sewer at frequent intervals throughout the period of flood flows. An estimate of discharge must also be obtained for each sample, using a staff gage and rating curve, or other means. Analysis of the samples for the pollutant involved gives concentrations, from which are derived loadings. The record of pollutant loadings throughout a given storm hydrograph allows estimation of the total loading of the pollutant for a given storm event.

In order to save laboratory time, the technique of flow-proportional composite samples may be employed. The one composite sample then represents a weighted mean of the various individual samples and may be used as representative of the entire storm event flow. It remains to estimate from several storm event loadings the corresponding loading for a critical period (such as a design storm) or for a complete year, if total annual pollution is critical. There are a number of approaches which can be taken, such as the STORM and SWMM models, no one of which has been proven entirely satisfactory. A method has been developed which requires determining total loadings throughout at least three storm events.[14] The researchers found that all pollutant storm event loadings have a tendency to increase with the peak flow of the storm, though not strictly proportionately. By plotting three or more points, a line can be drawn from which flows at any other storm can be estimated. See Figure 4-2.

If annual total or average loadings are required, it is necessary to use a rainfall-runoff relationship to convert rainfall to flow rates, for rains determined to be characteristic of the year and region, on the basis of which total loadings for the year could be derived.

Of course, storm characteristics other than peak flow may have an influence, so that inclusion in the record of nontypical storms could distort the predictions. Also, there may be seasonal differences. Research for a recent Rutgers doctoral dissertation indicates that on small suburban watersheds in New Jersey there are very considerable seasonal differences in the relationship of BOD pollution loading to discharge, but that for each season the BOD loading is closely proportional to stream flow.[15] (See Figures 4-3 and 4-4.) Although this analysis was conducted with individual observations rather than complete storm event loadings, the data basis was quite large. This relationship, if confirmed, would indicate that total runoff during a storm event, and not peak flow, would be proportional to storm event loadings, and that the analysis, to be valid, must be conducted on a seasonal basis.

Available data indicate that other pollutants do not usually occur proportionate to flow but with some other relationship.[16] It has been widely assumed, particularly in EPA's much advertised SWMM model, that nonpoint pollutants accumulate at a constant rate, and are washed off by rainfall in quantities which are proportionate to the length of time since the last storm. A considerable body of recent research in New Jersey indicates serious doubt as to this relationship.[17] No significant correlation was found between the total pollution loading

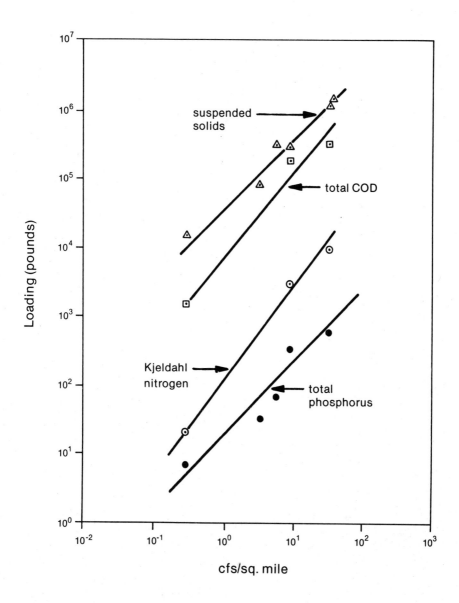

Source: Zogorski, T.S. et al. in "Urbanization and Water Quality Control," American Water Resources Association 1975. Copyright by the American Water Resources Association. Used with permission.

Figure 4-2. Pollutant Loadings in Storm Runoff, Panther Branch

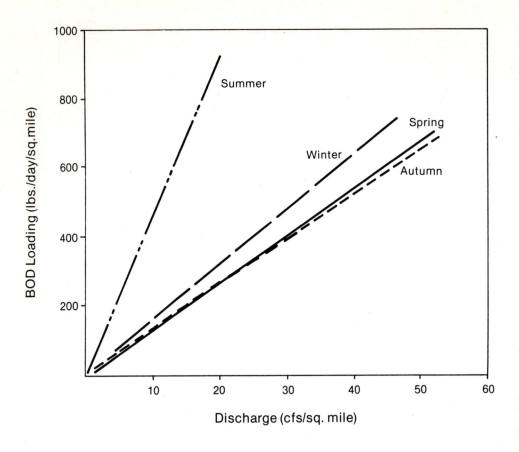

Source: Adapted from Donald V. Dunlap, Ph.D. thesis, Rutgers University, 1976.

Figure 4-3. Seasonal BOD-Discharge Relationships, Six Mile Run

of BOD or of heavy metals with the duration of the intervening period between storms. In principle, there should be some relationship, but apparently it is much less important than the characteristics of the rainfall and major seasonal influences. Obviously, pollution runoff models based upon such approaches should be used with great caution until improved methodology becomes available. Clearly, more research is indicated. The additional data as to storm event loadings now being evaluated by Section 208 studies should allow relatively rapid improvement of this situation.

Two possible sources of misunderstanding must be especially avoided. Where large combined sewer overflows occur, naturally the streams will have higher pollution, by an indeterminate amount, than that of similar streams carrying no sanitary sewage.

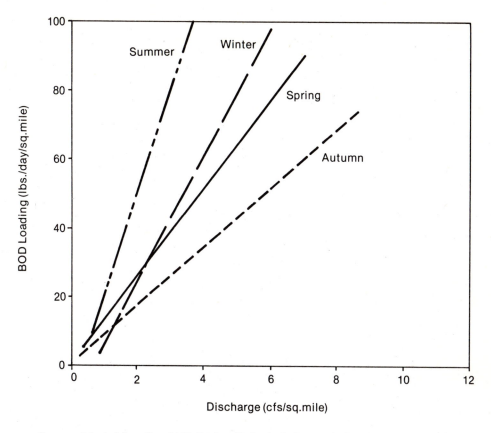

Source: Adapted from Donald V. Dunlap, Ph.D. thesis, Rutgers University, 1976.

Figure 4-4. Seasonal BOD-Discharge Relationships, Millstone River

Also, some data has been obtained without modeling, by simply subtracting inflow loading from outflow loading of a long stretch of river. Since BOD and ammonia particularly are changed with time, so as to be reduced in quantity, this procedure characteristically understates the amount which entered the streams.

BOD in Urban and Industrial Runoff

The Council on Environmental Quality in its annual report of 1974 emphasized the importance of storm water runoff as follows: "Until the storm water situation is analyzed and efficient corrective measures taken, there is little or no sense in seeking higher levels of treatment efficiency in existing secondary plants. In Roanoke, for example, removal of BOD was upgraded from eighty-six percent to ninety-three percent yet there was no dramatic reduction in the BOD load (3.2 million lbs in 1969 compared to 3.06 million in 1972)."

Since there has been no systematic national program for monitoring urban runoff pollution, no one has very good estimates of the loadings involved, except where research has developed an insight.

A study of the Delaware Estuary recently made by Enviro Control, Inc. for the Council of Environmental Quality indicated a characteristic decrease in DO after a storm, averaging 2.05 mg/l.[18] The estuary took an average of six days to reach the minimum DO following the storm, plus another two to six days to recover. This confirms the massive organic shock loads which are conveyed to the estuary, although these loadings remain to be estimated by direct means. It is apparent that urban runoff will greatly influence the water quality.

Table 4-1 (Pollutant Loadings from the Urban Watershed) shows data considered reasonably accurate from the monitoring and analysis of a number of urban and urbanizing watershed, of areas up to about ten square miles. It will be noted that there is a general tendency for increased BOD loadings in the table to be recorded for the areas of heavier urbanization, the cases of single family residential housing being much lower than other figures. Single family housing figures indicate a mean of about 29 lbs/mi^2/day, as compared to a range of 40 pounds to over 200 pounds for urban developments including substantial proportions of multiple family residences, commercial and industrial development. (For three of the listed areas, BOD data were not actually obtained; the figures given have been roughly approximated by the relationship to total organic carbon.)

The information contained in the table appears to be the best available; but it is quite limited in scope. It was obtained by a few researchers who have made serious attempts to include storm loadings in their data.

It is interesting to compare these figures to available BOD data from the NCWQ Report which indicated BOD loadings of 2-6 lbs/mi^2/day for farms.[19] Somewhat higher figures were obtained from mid-slope farms and wooded land in New Jersey (Table 4-2).[20] These show mean concentrations ranging from .9 to 2.2 mg/l and loadings of from 4.7 to 29.1 lbs/mi^2/day.

Colston and Tafuri have made probably the most intensive study of urban runoff in a small basin, in connection with an EPA-funded study.[21,22] The area in question is a 1.67 square mile heavily developed urban area in Durham, N.C. Thirty-six separate storm events were sampled during the year 1972, using an automatic sampler to take samples throughout the storm hydrograph. The annual pollutant yield was calculated from base flow periods and periods of urban runoff. For most parameters of pollution, the base flow provided only a small part of the total loading, for example five percent of the COD and eleven percent of the TOC. However, base flow produced fifty-four percent of the total Kjeldahl Nitrogen and twenty-four percent of the total phosphorus, suggesting the possibility that sewage leakage or other steady discharges may have been

Table 4-1
Pollution Loadings from Urban Watershed

Place	Size, Land, Use	Class	Loadings Pounds/mi² /day BOD₅	P	
Denver St.-Randall	.54 mi² , 95% single res.	Single res	17[a]	1.3	Incl. Storm hydrograph data
Lower Bull Run-Randall	6.5 mi² , 55% single res. 14% multi. res. 9% commercial	Heavy urban	21[a]	9.1	Incl. Storm hydrograph data
Durham-Bryan	1.7 mi²	Heavy urban	C.140	3.5	
Morristown	5.2 mi² , 39% residential 5% commercial 2% industrial	Medium urban	75.2[b]	46[b]	Incl. wet and high flow periods
Greenfield-deGiano	18 mi²	Light urban	163.5[c]	7.9[c]	Incl. wet and dry day data
Durham-Colston	1.67 mi² , 59% residential 19% comm. & industrial	Heavy urban	C.200[a]	8.2	Intensive storm hydrograph sampling
Beaverdam	0.5 mi² , 82% single res.	Single res.	33		
Bennington Parkway	0.3 mi² , 60% single res.	Single res.	33		
Mile Run	1.0 mi² , 38% res. (mainly multiple) 14% commercial 5% industrial	Heavy urban	125		Wet and dry day data
Portner Ave.-Randall	90% commercial	Medium urban	41		
General	Urban range		40-700	2.5-5.0	NCWQ Staff report

[a]Assuming that BOD₅ approx. 2/3 TOC. This is an approximation.
[b]This was for a very wet year. An average year would presumably have lower BOD loadings, possibly to 50 lbs/mi² /day.
[c]Storm sewer overflows increased loadings in wet weather.

involved. Estimated storm pollution concentrations and yields in pounds/mi² / day are as given in Table 4-3. Comparisons of total pollutant loadings measured in COD with the pollution from the sewage plant serving the same area, indicated that the urban runoff produced almost as high a COD loading as did the raw sewage, and several times as much COD as the effluent of the plant after treatment (ninety-one percent efficiency of removal). Heavy metals data also were obtained as discussed in the next section.

Table 4-2

BOD Loadings and Concentrations in Undeveloped Watersheds in New Jersey

Place	Size, Land, Use	Class	BOD Load lbs/mi² /day	BOD Concentrations mg/l
Six Mile Run	10.8 mi² , 65% farms & crops, 17% res.	Farmland	29.1	2.1
Rock Brook	5.9 mi² , 80% wooded 10% farms	Wooded	6.5	.9
Big Bear	9.0 mi² , 57% farms & crops, 13% res.	Farmland	4.7	2.2
Duck Pond	5.7 mi² , 60% wooded, 30% farms & crops	Wooded	11.1	2.0

Unfortunately, Colston encountered difficulties with the dilution techniques used in measuring BOD and was unable to obtain any BOD values in which he had confidence. Also, the location of his automatic samplers, near the bed of the stream, gave results which he found to be not entirely representative of the total stream flow, since solids tend to occur in larger concentrations in the lower portions of flow. However, there is no doubt that this part of EPA's research effort produced valuable results.

A completely different approach was adopted in a research project funded by the EPA in 1974, mentioned above, which had as its objectives both the development of technology and a nationwide survey of nonpoint source pollution throughout the United States. The agency's Request for Proposal issued to accomplish this ambitious project called for completion within a single year, without any data gathering whatsoever! The contract was awarded on the basis of estimating urban runoff indirectly from the pollution found in street sweepings. This EPA research indicated that street dust contains significant quantities of lead, zinc, nickel, chromium, copper and petroleum, with lead from this source particularly important.[23] However, it could not possibly estimate the "urban runoff" because street runoff is only one of many sources. (The term urban runoff is usually used to include all pollution emanating from a city except known point sources of untreated wastes and treatment plant effluents.)

Another approach which is more plausible is the estimation of urban runoff based upon routine sampling of medium sized urban or suburban streams. For reasons given under the heading of "Technology," above, this usually results in underestimating the pollution produced by a given watershed, and always results in underestimating the amount of nonconservative pollution flowing into the watershed. However, it may be argued that from the point of view of effect upon receiving waters below, the output loading is the relevant figure.

Thus, we have Ragan and Dieteman[24] reporting on BOD concentrations averaging less than 2.0 mg/l for each of three groups of streams in suburban

Table 4-3
Urban Runoff, Durham, N.C.

Mean mg/l Concentration		lbs/mi² /day
170	COD	530
42	TOC	326
205	Volatile solids	
1223	SS	
.96	Kjeldahl N as N	10.7
.82	Total phosph. as P	8.2

Note: Fecal coliforms averaged 23,000/100 ml

Maryland, which drained areas classified as entirely rural, up to twenty-five percent urbanized (developed) and over twenty-five percent urbanized (developed). Dissolved oxygen was high in each case, averaging over 10 mg/l for each group. BOD loadings averaged only 5.6 lbs/mi² /day for the areas over twenty-five percent developed, as compared to 4.5 pounds for the entirely rural areas. The data gathering was extensive, and it appears from the data that there is no BOD problem involved. The drainage areas analyzed consist largely of bedroom communities serving Washington, D.C., so that some of the areas over twenty-five percent urbanized probably have mainly single family housing and farm land, which previous studies have shown to be relatively low outputs of organic pollution.[25]

Also, the Maryland data consists of individual samples taken by routine sampling. Previous experience elsewhere has shown that, unless great attention is paid, routine sampling will not often be undertaken during storms, but will favor pleasant weather. Substantiating this is the fact that Ragan found that fish diversity was very markedly reduced in the more urban areas, which he attributed primarily to physical erosion of the streambeds, in spite of the low suspended solids figures quoted, averaging 9-18 mg/l. These results compare with data taken by Randall with automatic samplers at Lower Bull Run in Virginia, which showed mean suspended solids of 6 mg/l for base flows and 617 mg/l for storm events. For a residential area, Randall's figures showed suspended solids of 750 mg/l during storm runoff. If sediment pollution adversely affected the fish in Maryland, it probably was from storm flows, with sediment concentrations of hundreds of mg/l, which covered up the beds of streams and also affected the fish directly. However, other pollutants may have been involved. Environmental impacts of urban stormwater cannot readily be identified by data from routine water sampling of sizeable streams, particularly if drainage areas are only partly urbanized.

Increasingly, urbanization of the metropolitan hinterland takes place through large developments of multiple family housing and large shopping centers. Data quoted in tables above were derived from conventional older developments, in which the suburban housing was mostly single family units on small lots. Multiple family housing was confined to city streets, usually including commercial areas in the same watershed.

Unfortunately, some erroneous impressions have been created by a federally sponsored report in 1974 which purports to prove that multiple housing creates less pollution per housing unit than single family homes.[26] Analysis of the detailed cost analysis indicates that a basic assumption was made that pollution varied in direct proportion to runoff. Since there is less runoff per housing unit from densely packed units, it was concluded and reported that there would be also less runoff. This grossly unscientific conclusion was apparently not detected by the federal agencies concerned, since this aspect of the report was only a side issue.

Hard data are lacking, but the probability is that the more densely packed housing units will create more pollution in runoff, since waste materials will flow more directly into storm drainage, rather than being washed over and absorbed by lawns and other vegetation.

Heavy Metals

Wilber and Hunter have monitored heavy metals from a heavily urbanized watershed at Lodi, New Jersey, 2.2 square miles in areas, and almost entirely developed.[27] Of the two subbasins, one had twelve percent industrial use and fifty-eight percent commercial use and the other five percent industrial and forty-two percent commercial. Samples were taken manually at five and 10 minute intervals throughout seven storm events, giving an unusually complete basis for analysis of storm runoff. After initial studies to characterize occurrence, flow-weighted mean average concentrations were used to determine total loadings for a storm event. The major heavy metals found were lead, zinc and copper, and the highest concentrations usually occurred within the first thirty minutes of storm runoff, showing a very pronounced first flush effect. See Figure 4-5. For large rainfall, the loadings increased much more than proportionately to flow. (Figure 4-6). It should be noted that the one extraordinarily high concentration of nickel shown for a low rainfall unquestionably represents a dumping of polluted material.

Comparison of mean heavy metals concentrations in urban stormwater at Lodi with concentrations reported elsewhere are shown in Table 4-4.

These concentrations are many times higher than heavy metals concentrations in medium streams as shown in the EPA annual report for 1975.[28]

Analysis of urban runoff pollution by University of Massachusetts researchers showed a particularly significant relationship of pollution from heavy metals

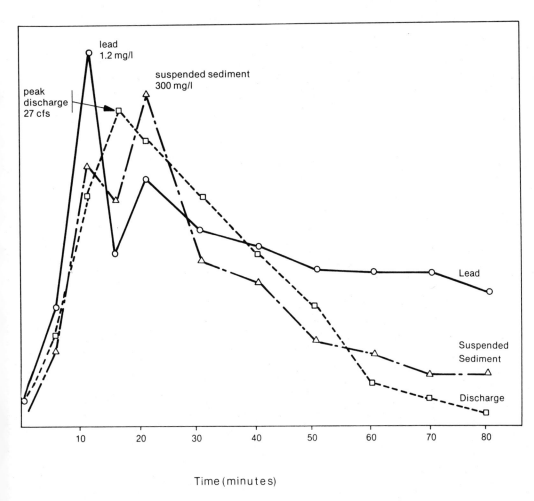

Figure 4-5. Storm Runoff of Sediment and Heavy Metals

to the condition of the benthic macroinvertebrate community, the small creatures inhabiting the bottoms of streams, which constitute such an important part of the food chain for fish.[29] At the upper stations, larval forms of mayflies, stoneflies, caddisflies and midges were found in abundance. Species diversities, particularly by late summer, were much reduced at stations downstream in the urban area. In fact, at the lowest station, snails comprised ninety-nine percent of the total population over the last six months observed. Obviously, some adverse influence eliminated or impeded colonization of the bottom by various species. Initial attempts to link this diversity to changes in regularly monitored physical and chemical parameters were unsuccessful, since minnows and dace were

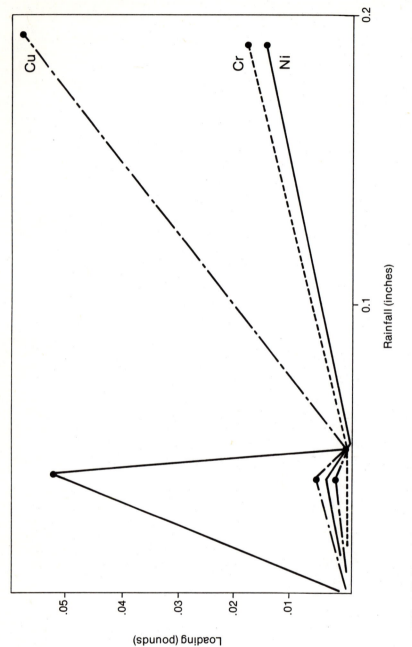

Rainfall (inches)

Loading (pounds)

Cu

Cr

Ni

0.1

0.2

.05

.04

.03

.02

.01

Source: Wilbur, W.G., and Hunter, J.V. "Urbanization and Water Quality Control," American Water Resources Association, 1975. Copyright by the American Water Resources Association. Used with permission.

Figure 4-6. Storm Event Loading of Heavy Metals

Table 4-4
Concentrations in Stormwater in Lodi, N.J.

Metal	Durham N.C.	New York N.Y.	Palo Alto Cal.	Okla. City Okla.	Lodi N.J.	NCWQ Staff Figures (urban)
Lead	.46	–	.093	.23	.87	–
Zinc	.36	1.6	–	–	.47	1.24
Copper	.15	.46	–	–	.18	–
Chromium	.23	.16	–	–	.06	–
Nickel	.15	.15	–	–	.03	–
Reference	(30)	(100)	(88)	(135)	(115)	(179)

continuously in evidence, and the levels of metals in the water column were negligible at all stations. However, it was found that metals concentrations in the sediment and detritus were greater at the lower stations, and the benthic organisms contained metals at still higher levels, ranging to two orders of magnitude greater than normally accepted acute toxicity thresholds. This explanation fits neatly into studies of heavy metals by Wilber and Hunter,[30,31] which showed the occurrence of heavy metals very largely concentrated in the sediments of the first flush of storm runoff.

The mean loadings of metals in the stream at Lodi were estimated on the basis of the computed annual runoff, with the assumption that the flow-weighted mean concentration observed would apply. Loadings for basins of different land use were as shown in Table 4-5, compared to results of Colston in Durham.

In considering environmental impacts of heavy metals, it should not be assumed that the amounts observed will move uninterruptedly down the stream. On the contrary, most of the lead and zinc at Lodi occurred in particulate form, and might be expected to be deposited as sediments upon arriving in a larger stream if the latter were not in flood.[32]

Analysis was also made of the heavy metals in rainfall at Lodi during twenty-four precipitation events. However the computed amounts of metals in rainfall for each metal was an insignificant proportion of the total loading from stormwater runoff.

Illegal dumping often takes place during storms. Data shown in Figure 4-7 can only be accounted for if heavy dumping of polluted materials occurred fifty minutes after the storm began. Such results indicate that the pollutional significance of stormwater runoff cannot be even approximated by normal monitoring methods, and absolutely requires analysis of complete storm hydrographs. The usual BOD modeling based on low flow conditions would give no clue to the pollution occurring at either Greenfield, or Lodi, nor would even routine monitoring of metals concentration in the streams.

Table 4-5
Heavy Metals Loadings from Urban Runoff

| | | Loading lbs/mi² /day | | |
| | Durham, N.C. | Lodi, N.J. | | |
Metal		Industrial/ Commercial	Residential/ Commercial	Residential
Lead	5.1	14.7	7.6	6.4
Zinc	3.5	7.1	4.9	3.2
Copper	2.8	14.9	2.3	1.5
Nickel	2.1	.69	.66	.55
Chromium	2.8	.29	.29	.35

The above scanty data on occurrence of heavy metals in urban runoff is sufficient to run up a red flag, but not sufficient to apply the data elsewhere with any degree of confidence. There was a great deal of variation in the data. On one occasion at Lodi a stormwater sample showed 32 mg/l of copper which must have related to an unauthorized release of considerable magnitude. The correlation of heavy metals occurrence with population and general economic development will probably be less feasible than for BOD. Lead probably relates partly to volume of automotive traffic, but all of the metals are strongly related to industrial activities, and in many cases probably originate in specific practices which should be eliminated.

Nutrients

In the usual concentrations in which they are found in streams, nutrients are not toxic to natural species. Their adverse effects relate to the eutrophication of lakes and impoundments, and to a much lesser extent, production of algal blooms in rivers and estuaries, as discussed in Chapter 3. Eutrophication and algal blooms in inland waters are usually dependent upon the quantity of phosphates present, since nitrates are usually present in more than sufficient quantities.[33] Therefore, it is with phosphates that nutrient planning must be particularly concerned, except for the special problem of nitrates in drinking water.

Phosphorus loads, like other nonpoint pollutants, are not adequately represented by monitoring of low flow conditions. A single storm event may contribute a total phosphorus load equivalent to a thirty day low flow, steady state loading.[34] The concentration of total phosphorus increases markedly after heavy rainfalls, and consequently the storm loadings are extremely large. The soluble reactive phosphorus, however, does not increase nearly as rapidly with flow in the streams studied. See Figure 4-8.

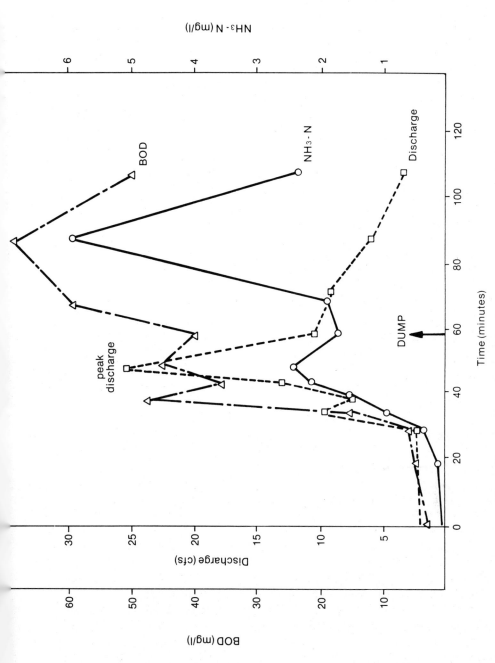

Figure 4-7. Storm Event Pollutant Concentrations, with Dumping

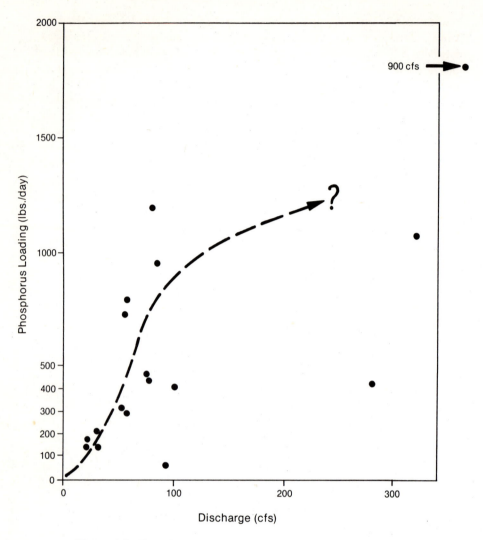

Figure 4-8. Phosphorous Loading and Discharge, Saddle River

Randall and associates found that urbanized areas showed very large loadings of nutrients.[35] In the urban area, twenty-four percent of the total phosphorus was contributed by storm flows, compared to one-half of the total organic carbon, and ninety-eight percent of the suspended solids.

Unit loadings of total phosphorus for various urban areas and one residential area are given in Table 4-1. Most of the loadings for urban areas fall in the range of 5 to 9 lbs/mi[2]/day. Uttomark quotes other data for total phosphorus from urban areas of 4.9 lbs/mi[2]/day (Ann Arbor urban) and 1.7 lbs/mi[2]/day (Madison residential).[36] The EPA annual report for 1975 indicates that the Missouri, Upper Mississippi and Yazoo rivers have total phosphorus concentrations averaging 25-30 mg/l, whereas the Columbia River and most Eastern rivers cited averages below 10.[37] Obviously these are large regional differences, therefore nationally derived averages should be applied with caution.

In rural and residential areas, substantial amounts of nutrients originate in commercial fertilizers. The action of phosphates and of nitrates is quite different. The phosphorus in fertilizers combines with clays and remains locked in the soil until either it is utilized by plant life or the soil erodes, in which case the phosphorus flows with it as suspended sediment. Soluble phosphorus on the other hand is more apt to occur in sewage or other organic material.

Nitrates in the soil remain much more soluble. If fertilizers are used in proper quantity, the nitrates move slowly with soil moisture but are largely used by plants during the growing season. During the late winter and occasionally in midseason, following exceptionally heavy rainfall, nitrates may pass below the root zone into the groundwater.[38] This is not without its dangers, the consequences of excessive nitrates being said to include occurrence of "blue babies." Nitrates are not allowed in drinking water in excess of ten parts per million under the new Clean Water Act. In Southern Illinois, groundwater with excessive nitrates was found in wells, but the pollution was manifested not by human illness but by serious effects upon infant pigs. Standards for drinking water are rightfully set conservatively, because we cannot carry out research with human subjects to determine the risks of given conditions.

Uttomark finds that total phosphorus produced from forested areas varies from .02 to 1.35 lbs/mi^2/day, with higher values generally from sloping lands.[39] He reports that agricultural lands produce from .05 to 3.6 lbs/mi^2/day. The NCWQ staff estimated .5 to 1.0 lbs/mi^2/day from agricultural lands. The EPA report for 1975 reports mean loadings of .23 lbs/mi^2/day from forest and .48 lbs from agricultural lands.[40] The general picture is quite clear. Undeveloped lands produce relatively little phosphorus, agricultural and residential areas more, and general urban developments still more phosphorus for a given area. (This is, of course, exclusive of point sources such as waste treatment plant effluents.) For this reason, it is apparent that the eutrophication process in lakes will be greatly influenced by the extent of development of the surrounding watershed.

Considerable information is becoming available as to nitrate loadings and concentrations coming from various areas. These data must be interpreted with great care, because of the nitrification process, which is most active in small, shallow, rocky streams, and denitrification, which occurs in sediments. The nitrogen in a stream, besides changing form, may either increase or decrease in total amount, depending upon a complex set of circumstances. Besides, as indicated in the next chapter, control of total nitrates may not be a practical approach to limit eutrophication, and among the nitrogen forms only the toxicity and oxygen demand of ammonia may be important.

Hydrocarbons

Although some classes of hydrocarbons are potentially extremely toxic or odoriferous, and most of them are large consumers of oxygen, it is only recently that information has begun to be developed about the occurrence and characteristics of hydrocarbons in wastewater and urban runoff. The usual measure of organic pollution, five day BOD, does not correctly indicate the oxygen demand of petroleum products, because they are only slowly biodegradable. Moreover, hydrocarbon chemistry is so complex that the test for oil and grease has usually been applied instead, which includes oils and fats from kitchen wastes and sewage as well as hydrocarbons.

However, recently Hunter and associates measured petroleum hydrocarbons in stormwater samples throughout the hydrographs of three storm events, for a 13-foot storm sewer in North Philadelphia.[41] This storm sewer drains an area of 2.4 square miles. Land use characteristics were as shown in Table 4-6.

This appears to be the first reporting on petroleum hydrocarbon content in urban storm waters.

The mean concentration of hydrocarbons during the three storms ranged from 2.18 mg/l to 4.04 mg/l, with no obvious relationship between runoff volume of each storm and the mean concentration. Most of the hydrocarbons, about seventy-nine percent of the total, were associated with particulates. The occurrence during the storm indicated a fairly marked first flush characteristic, as indicated in Figure 4-9. Flows during dry weather appeared to have no appreciable petroleum loading and were disregarded.

Further investigation is proceeding under a project sponsored by NSF-RANN. In addition to estimating quantities of petroleum entering the estuary from storm runoff, municipal waste treatment plants, and industrial wastewaters, it will be necessary to characterize the vast number of hydrocarbons involved, determine how they may be affected by chlorination, and what their effect is upon the environment. (Effluents from waste treatment plants are almost always chlorinated and the EPA has recommended disinfection of storm

Table 4-6
Land Use Drained by North Philadelphia Storm Sewer

Land Use	Part of Area
Single family residential	20%
Multiple family residential	32%
Industrial	1%
Commercial	29%
Public and quasi-public	7%
Conservation, etc.	11%

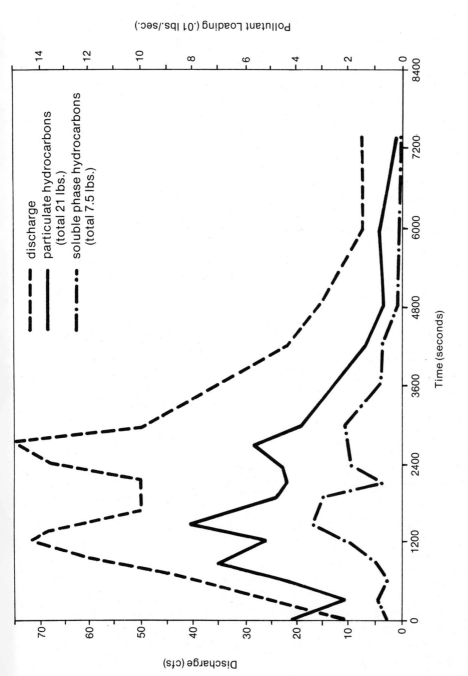

Figure 4-9. Petroleum Hydrocarbons in Urban Runoff

runoff, if it should be treated. Some chlorinated hydrocarbons are potent and persistent pesticides, and others are believed to be carcinogenic.) It would be very appropriate and useful if EPA would fund this sort of research. These questions must be answered before anyone can tell whether the current and anticipated future regulation of the petroleum industry will hit the mark or miss it entirely, and whether much better use of available resources in remedying petroleum damage to the environment might not be made.

Combined Sewers

The majority of the older sewerage systems of the United States use combined sewers, which collect some or all of the storm runoff as well as sanitary wastes. After heavy rainfall, these combined sewers convey quantities of polluted water, far in excess of the capacity of the waste treatment plants; therefore some kind of a flume or diversion gate is provided, which bypasses part of the flow directly to the receiving waters. The urban runoff carried by the sewers in time of storm is highly polluted itself, and the part diverted into receiving waters is also mixed with raw sewage, therefore, a combined sewer storm overflow constitutes a major source of pollution.

As an example, a recent study in Chester, Pa., showed that combined sewer overflows amount to one and one-half times as much BOD loading as the effluent from the secondary waste treatment plant serving the same area.[42] Obviously, if the average rate of pollution was this high, it must have been several times as great during heavy rainfall.

From data given in the Chester study, it appears that the regulators were set so as to reduce the flow to the waste treatment plant during storms to less than half the normal flow, so it may be surmised that the loading of the combined sewer storm overflows could have been largely reduced by relatively simple adjustments. However, it is routine for combined sewers to operate improperly. The Interstate Sanitation Commission in 1972 studied ten combined sewer overflow systems in northern New Jersey and New York, and concluded that over twenty-five percent of the regulators were inoperable, that inadequate maintenance was responsible for large quantities of overflow, and that data as to operating conditions were generally either inadequate or incomplete. Some combined sewers overflow and pollute receiving waters even during dry weather, and others every time it rains. Other better designed systems, having higher treatment plant capacity, may overflow only two or three times a year.

There are from fifteen to twenty thousand diversion structures discharging waste from combined sewers in the United States. It is estimated by an EPA spokesman that it would cost $30 billion to replace these combined sewers with separate sewers carrying the storm runoff and sanitary sewage separately. A few years back enthusiasts and most leading pollution control authorities were sure that all combined sewer systems should be replaced with separate systems for stormwater and sanitary wastes at an early date. However, the enormous costs involved in digging up the congested streets of many of our older cities, and the dubious effect upon receiving waters, has caused a much more restrained view to be taken. Another important consideration is the fact that the stormwater runoff (which includes all sorts of miscellaneous spills, seeps, leakages and illegal discharges as well as runoff proper) is itself so highly polluted that in some cases it may have to be treated, in which case the expense of providing two separate sewer systems would be wasteful.

Once again we see that quickly-devised, universal solutions to water quality problems, formulated with "bandwagon" enthusiasm, and sensitive to the latest headlines, are not the best ways to arrive at optimum solutions. Traditional combined sewers constitute a serious pollution problem, true, but careful study will be required in each case, covering the entire water and water pollution situation, to decide how this problem should be solved.

Lawns and Gardens

Upper middle class dwellers in suburbia are characteristically among the most consistent and ardent supporters of the environmental movement, and most of them would certainly be surprised to be told that their carefully tended neighborhoods are a source of pollution (over and above the septic tanks which are used in most delightful country residences). However, most suburban lawns are apt to be over-fertilized, and often over-watered; moreover, the lawns may be carried down to the very edge of lakes and small streams. Although statistics are not available, there is no doubt that some of the eutrophication in small ornamental ponds and in lakes further downstream is due to this source. What is required for control is prevention of erosion, and minimum use of water and fertilizer (particularly within fifty feet of a stream). Heavy fertilization in the fall is counterindicated.

Control of Urban Runoff Pollution

It is difficult to say how many installations for improving storm water pollution exist in the United States, because the very concept is a new one. A 1972 survey of the American Public Works Association[43] showed that 1400 on-site runoff detention facilities are operational in about 100 local jurisdictions reporting, with special mention of the Nassau County, N.Y., program of stormwater determination basins, and the increasing number of basins of the Los Angeles area. However, both of these major programs relate primarily to ground water recharge, and most of the others undoubtedly are used in aid of flood retardation. Since there is no regular federal program in aid of urban runoff pollution control, it is only in exceptional circumstances that cities have undertaken to construct such facilities, or to plan them in detail.

San Francisco

Alan Friedland of San Francisco explains a cost-effective control system planned for urban runoff in that city, which is of considerable interest.[44],[45] The master plan was designed to improve the water quality for the city's harbor and beaches. It was based upon data from a very complete monitoring system established by the city.

First considered was a sewer separation plan for San Francisco's 900 miles of sewers. The cost was estimated at three billion dollars; moreover, it would do nothing to control urban runoff pollution and would create unimaginable disruption of city streets.

The second proposal was to construct fifteen to eighteen large treatment plants at various shoreline basins, which similarly would cost roughly three billion dollars, with major logistical problems and high operating costs.

The third proposal was to greatly enlarge the three existing sewage treatment plants and provide storage to control runoff. This was estimated to cost about two billion dollars.

Lastly, after some very careful study of the actual hydrological and pollution situation, a plan was worked out involving nine million cubic feet of inland and shoreline storage, a 100 mgd chemical and primary wet weather treatment plant, a five mile ocean outfall, and fifteen miles of tunnel or transport main lines. The cost was estimated at something over $400 million. This is a very large sum, but only about one seventh of the cost of the alternatives first considered. The plan has not yet been built; but at least the program is known.

Friedland found that there was determined public opposition to large detention basins, so that large covered channels had to be planned instead. The new channel and pumping arrangements would reduce combined sewer overflows to receiving waters from an average of eighty-two to eight per year.

Chicago

The Chicago area has had a similar problem. In this case, the city and adjacent communities are not allowed to discharge any wastes at all into Lake Michigan, but must convey it by canals to the Illinois River, the quality of which is strictly controlled. The Metropolitan Sanitary District has set up monitoring and analysis programs to work out the best plan.[46] However, in this case, the integration of flood control and water pollution has been unusually difficult. Moreover, alternatives have been required involving deep-tunnel detention-storage schemes, which are technically novel and somewhat controversial. Gradually the details of the plan have been developed. It will involve large tunnels (or deep, open pits) into which urban runoff and combined sewer overflows will be channeled. These storage sites will provide interim aeration until the flows can be processed through the treatment plants. A very high degree of waste treatment (although less than the maximum) will be required in the municipal treatment plants. In addition, in order to meet water quality oxygen standards, a system of instream aerators or oxygenators will be needed. The EPA has advised that under PL 92-500 the aerators cannot be funded federally, even though they are admittedly the most economical means of achieving the desired goals. Estimates showed that $100 million invested in river aeration would save an additional expenditure of over $250 million additional costs for still higher degrees of advanced waste treatment.

No one can visualize just how the Corps of Engineers and the EPA, with their vastly different planning and funding approaches, can ever cooperate to build a system of tunnel storage such as visualized here. Moreover, it is not clear how the mandate of PL 92-500 to provide "best practicable" treatment can be reconciled with an approach which combines a lesser degree of treatment with other means to approach the desired goal. Nonetheless, as a result of prolonged and strenuous efforts by the Metropolitan Sanitary District of Chicago, a sensible plan is being devised, and hopefully the EPA and the state will find means to implement the optimum plan rather than insisting on much more expensive alternatives derived from the literal interpretation of a statute.

A Major Boston Problem

Pisano reports on a thorough study of alternative programs for reducing fecal contamination of five recreational beaches from the two high density residential communities of Dorchester and South Boston, Mass.[47] The original program centered on collection and conveyance of combined and storm sewer overflows to a central pumping plant, where the wastes were to be lifted and transported under pressure to a terminal location for detention, chlorination, and ultimate discharge into Boston's inner harbor. The estimated capital costs exceeded $100 million, and time of construction was estimated at five to eight years, with considerable community disruption.

The consultants applied a cost-effective approach focused on maximizing potentialities of the existing sewer system, and examined various new alternatives, including: 1) removal of deposits from an existing interceptor, 2) eliminating seawater intrusion into it, 3) upgrading inoperative control structures and cleaning plugged conduits, 4) daily flushing with water of critical sections of the line, 5) off-line storage facilities, and 6) chlorinating at selected overflow points.

As a result of these studies, a whole range of alternative combinations was evaluated. Figure 4-10 shows variations of BOD discharge as a result of cleaning, flushing and storage. Flushing programs A, B and C shown denote, respectively, no flushing, a low level flushing program for 100 segments, and a high level flushing program for 420 segments. Similar relationships for sediment discharge were derived, and water quality responses for the various alternatives were evaluated.

It was finally found that a program equally as effective as the original plan could be obtained by cleaning the present interceptor and preventing seawater intrusion, creating eighteen million gallons of off-line storage in Dorchester, utilizing the low level flushing program, and chlorinating certain overflows. The total cost was evaluated at $38 million, compared to $100 million for the capital costs only of the original program.

Obviously the above constitutes a success story highly creditable to the local planners and the consultants involved. However, from a policy viewpoint, other aspects are important. In the first place, the successful later planning did not start with a formula or rule dictated from Washington, but with an impartial study of all alternatives, including a desired environmental objective. It is only through this basic approach that optimum solutions can be obtained.

Urban Stormwater Treatment Funded by the EPA[48]

Under the categories of research, development and demonstration, the EPA has funded or aided in the funding of substantial efforts to improve the technology of treatment of urban stormwater (as compared to EPA neglect of planning and environmental impacts aspects). The Metropolitan District Commission of Boston began operating the Cottage Farm Stormwater Treatment Station in 1971, as a detention and chlorination facility. Suspended solids were forty percent removed, and disinfection provided practically 100 percent reduction in coliform count. Results of treatment in reducing BOD were said to be erratic, showing little or no improvement. (Chlorination ordinarily drastically reduces indicated BODs, by killing most of the bacteria which use the oxygen, even though the organic content of the water may not have been materially affected. However, by dechlorinating and using bacterially seeded dilution water, results indicative of organic content can be obtained.)

At Racine, Wisconsin, two screening/dissolved air flotation tanks were constructed and began operating in 1973, for treatment of combined sewer overflows. The units were projected to remove sixty to seventy percent of the BOD and seventy to eighty percent of the suspended solids. Operating difficulties were encountered initially and actual results are not available.

The dual use of wet-weather treatment facilities in a major supportive role to normal sewerage plant operations is illustrated at a New Providence installation. Two trickling filters are installed, which are normally operated in tandem (flow first through one, then the other). When storm conditions increase flows

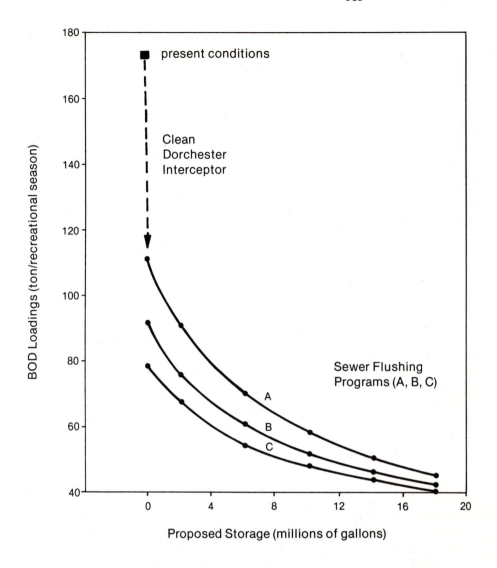

Source: Pisano, W.C., "Urbanization and Water Quality Control," American Water Resources Association, 1975. Copyright by the American Water Resources Association. Used with permission.

Figure 4-10. BOD Loadings for Alternative Control Programs

(mainly by infiltration into previous sewer lines) the two trickling filters are operated in parallel. This in effect doubles the capacity of the plant. The experiment was considered reasonably successful, although some loss of efficiency was noted at high flows.

At Mt. Clemens, Michigan, combined sewer overflows from a 212 acre test area are treated in three successive lagoons. The first lagoon acts as a combination and aeration basin. The second lagoon, separated from the first by a microstrainer, acts as an oxidation pond, and the third as an aeration lagoon. Effluent from the final lagoon is discharged after pressure filtration and chlorination. The second and third lagoons are being developed for recreational use. Removal of BOD and SS averaged better than ninety percent.

State-of-the-Art for Treatment of Stormwater

A good analysis of potentialities for treatment of stormwater is given in a recent paper by Field and Lager, mainly representing results of EPA-funded research.[49] Significantly, they found that there is no such thing as an "average" design condition, on account of the extreme intermittency and variability of storm-water runoff and related system flows; integrated systems of control and treatment are found to be the most promising. Main points of the summary follow.

Source controls may include flow attenuation, erosion control, restrictions on chemical use, and improved sanitation practices.

Collection system controls include flushing, inflow-infiltration control, improved regulator devices, sewer separation, and remote monitoring/control systems. A new device called the swirl concentrator offers the potential for effective solids concentration over wide ranges of flow, but the ultimate disposition of the solids is a major problem.

Storage is very helpful, but costly, and it requires valuable space, unless entirely underground. Treatment processes considered include sedimentation, dissolved air flotation, screening and filtration. Biological treatment plants might be successful if operated conjunctively with normal sewage plants, in order that the necessary biomass be provided by sludge.[a] Physical-chemical treatment of storm runoff is a possibility, but costs are high, and there is an increased mass of sludge to be disposed of.

Disinfectants are considered necessary, and the research program has used chlorine, sodium hypochlorite, chlorine dioxide and ozone. In practice chlorine is usually used (in the United States), although as noted above there may be some environmental hazards involved in the practice.

Corps of Engineers Approaches

Although the EPA itself has not engaged in planning water pollution control, the Corps of Engineers has. Regardless of the appropriateness of this institutional

---

[a]Rapid biological assimilation of organic wastes requires an adequate supply of the right kinds of bacteria, immediately available, which is furnished from recently-produced sludge.

anomaly, the planning has been vigorously done, and is summarized in more detail in Chapter 6.

The Corps of Engineers study in Seattle has evaluated the feasibility of alternatives to treating urban runoff, including land use restrictions and runoff controls. In various studies estimates have shown that costs of treating storm-water would vary from 25 percent to 125 percent over the cost of treating municipal and industrial wastewaters in the same communities.[50]

The Corps considers that the biggest hurdles in solving urban runoff problems may be the existing institutional arrangements. Drainage management is generally a piecemeal and often badly neglected program; no other element of our urban support systems receives so little attention in terms of regulation, supervision and financing. Traditional methods are used almost exclusively and there are few large aggregations of talent in the municipal drainage field. Accordingly, it is difficult to see just how a program of treatment of urban runoff could be implemented under our present institutions, even if it was found to be desirable.

Optimizing Costs Between Storage and Treatment

As shown by Shubinski, optimization studies indicate clear relationships between three variables: (1) the rate of overflow to the receiving waters of untreated stormwater, (2) the capacity provided by treatment plants, and (3) the amount of retention storage provided.[51] Using conditions from the San Francisco system, with certain hypothetical data added, he has found least-cost combinations of storage and rates of treatment to meet given overflow criteria. It is interesting to note that a system averaging one overflow annually of untreated runoff costs about fifty percent more than one averaging five overflows annually. See Figure 4-11.

Comparison of Urban Runoff Treatment with River
Oxygenation, Delaware River

A recent study considered various methods of reducing urban storm water loads, including a variety of methods based either upon separation of combined sewers or temporary storage, in conjunction with primary treatment, aerated lagoons, etc.[52] Experiences of a number of cities were tabulated. The variations in effectiveness and cost were very wide.

The unit costs cited were generally very high, especially where a considerable degree of effectiveness was achieved. Table 4-7 gives costs in various cities for the temporary storage of storm runoff followed by primary treatment to reduce BOD. The cheapest method indicated for removing BOD at an effective-

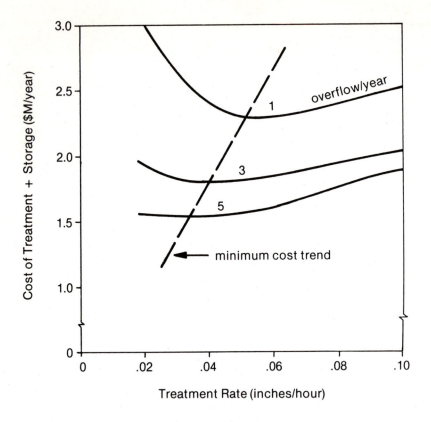

Source: Shubinski, R.P. "Stormwater Treatment vs. Storage," in Management of Urban Storm Runoff," Technical Memo No. 24, ASCE Urban Water Resources Research Program. American Society of Civil Engineers, May 1974.

Figure 4-11. Combined Sewers, Cost-overflow Relationship

ness over eight percent was $0.42/lb. However, BOD in levels characteristic of the Delaware is not harmful in itself; the main objective of lowering it is to maintain oxygen levels in the receiving waters. The dissolved oxygen levels may be maintained in large rivers by other means on a much more economical basis, as indicated below.

Studies and field tests of artificial aeration on the Delaware River have shown that with a continuous summer operation, the last milligram per liter of dissolved oxygen could be added by mechanical and diffuser aerators at about one-quarter the cost of incremental costs of waste treatment to achieve the same

Table 4-7
Cost of Reducing BOD in Storm Runoff

% Reduction	Unit cost/lb. BOD
8%	$ 0.12
48%	3.00
10%	13.00
22%	0.42
21%	2.59

end. Furthermore, subsequent research showed that for such a river, a mobile oxygenating craft could reduce costs still further. This study showed that a mobile oxygenating craft dispersing 400,000 pounds of oxygen daily could be built and operated on the Delaware Estuary for total costs including amortization of about $1,300,000 annually. For urban runoff, such a craft would only be operated about one-fourth of the time, at most, which would reduce total costs by about thirty-five percent. If transfer efficiency were seventy-five percent, and the craft operated thirty days of a five month period, oxygen imparted to the river would be 9,000,000 pounds and unit costs would be about $.09/lb., based upon 1972 costs. This is probably the cheapest way of adding additional oxygen to a large navigable river.[53,54,55]

Conclusions

The methodology of treating urban runoff and combined sewer overflows, and of combining such treatment with detention storage, remains to be perfected as far as BOD and suspended solids is concerned, and has not really been seriously considered as far as heavy metals and other pollutants are concerned. Methods of incorporating these elements in planning will be improved by the Section 208 studies, if they are done properly, but the lack of data will prevent definitive accomplishments in most cases. The devise of institutional approaches, including the reconciliation of the 1972 Act's strategy to the realities of the urban runoff situation, seems not even to have been initiated. It is quite obvious that desired water quality control cannot be achieved in most metropolitan areas without some form of treatment of urban runoff, but that costs are apt to be extremely high unless plans are really well-made. It seems highly unlikely that one formula, tailored in Washington, can do the job. The probability is that the Section 208 studies will mostly produce some hasty improvisations, based upon only a few months of data gathering, and that definitive planning will have to be done later, after a system of monitoring has provided a better factual base in each area, and additional research has improved available methodology.

5 Municipal and Industrial Wastewater Treatment

As brought out previously, treatment is the primary, almost the exclusive theme of our national water quality control program. However, the water pollution abuses were gross, and the improvement of water quality through direct treatment of wastewater had to be the first and primary approach to the pollution problems.

From a technological viewpoint, waste treatment was developed many years ago as the principal concern of sanitary engineering and as a standard aspect of engineering practice, at least as far as treatment of municipal sewage is concerned. Primary and secondary treatment in sewerage plants are familiar operations, and many of the concepts, equipment, and techniques worked out a generation ago and reflected in standard texts are still valid today, and are familiar to the profession. However, the term sanitary engineering has fallen into disfavor, and instead the euphemism environmental engineering has come to be generally used by engineers which may lead to confusion, as the study of environmental impacts and of instream conditions of pollution is also properly referred to as environmental science or environmental engineering.

The EPA has devoted a major part of its research effort to improving the technology of waste treatment and a great deal of information is now available.

The approach of present pollution control legislation is simplistic. It can be viewed as a moral injunction —"pollution is bad; therefore let us abolish it." Also the approach undoubtedly springs somewhat from the elemental aversion that we instinctively feel in the presence of human excrement. However, such feelings are not a sufficient guide to policy in such a complicated matter. Wastewaters and the various substances which they contain must be viewed from their effects upon the environment, and treatment processes must consider technologies and costs adaptable to a variety of situations, not simply the removal of all foreign substances to the maximum extent practicable. Therefore, this chapter includes summary information on various types of treatment, including disposition of sludges which treatment creates, also the disinfection of resulting effluents. State and federal requirements are reviewed briefly, and finally there is a summary of problems remaining.

Processes of Wastewater Treatment

BOD and suspended sediment can be removed from municipal sewage by secondary treatment, with predetermined effects and known limits of cost.

111

Usually in the United States this is accomplished by the activated sludge process, however, the effects of standard treatment processes on other pollutants are much more variable. Other processes are now being developed to meet the higher effluent standards required under the new law.

Bovet has given an excellent summary of treatment processes, which he classifies as follows:[1]

A. Preliminary treatment
   1. Screening
   2. Grinding
   3. Grit, grease and scum removal
B. Primary treatment (mechanical)
   1. Sedimentation in settling basins
   2. Mechanical aeration
   3. Final sedimentation
   4. Chlorination or other disinfection (if not followed by secondary treatment)
C. Secondary Treatment
   1. Trickling filter process
   2. Activated sludge process
   3. Digestion processes, stabilization pond or anaerobic digestion
      Note: Well-operated secondary treatment plants can usually remove eighty-five-ninety percent of BOD and over ninety percent of suspended solids.
D. Tertiary treatment (advanced waste treatment)
   1. Microscreening
   2. Lime clarification sequence
      a. Coagulation and flocculation and sedimentation
      b. Ammonia air-stripping
      c. Multi-media filtration
      d. Granular carbon adsorption
E. Nitrification and denitrification
F. Desalination

Effectiveness of certain kinds of tertiary or advanced waste treatment is given in Table 5-1, with respect to removal of pollutants indicated. Eiler for the EPA cited costs in 1970 of various chains of processes summarized as follows:[2]

List of Processes

1. Primary sediment and sludge disposal
2. Primary, activated sludge and sludge disposal
3. Miscroscreening

Table 5-1
Effectiveness of Various Processes

	SS	BOD	COD	TKN[a]	P	TDS
Microscreening	97.3	96	90	43.3	30.8	–
Lime Clarification Sequence:						
Step a	93.5	98	93.4	50	46.2	–
Steps b-d	99.1	99.5	98.6	90	84.6	–

[a]Total Kjeldahl nitrogen, including NH_3 and organic nitrogen but excluding nitrite and nitrate nitrogen.

4. Single stage lime clarification
5. Two stage lime clarification
6. Ammonia stripping and recarbonation
7. Multi-media filtration
8. Granular carbon adsorption (40 min.)
9. Chlorination

In view of technological change and continued currency depreciation, these costs are only roughly indicative of what could be expected today; but they do illustrate the sharply increasing incremental costs of removing given quantities of pollutant for the more refined stages of treatment (see Table 5-2). They also illustrate how meaningless it is to try to decide whether one of the process chains represents "best practicable" treatment.

Phosphorus

If phosphorus is to be removed from wastewaters, information is available as to the costs involved as shown in Table 5-3.[3]

Industrial Wastes

Treatment of industrial wastes is much more complex than treating domestic sewage, and the technology is not as well known. Although municipalities handle much industrial waste, the more specialized approaches to industrial waste treatment are practiced largely within industry. Detailed information as to processes and results, especially with regard to the relationship of production costs to amounts of waste produced, is frequently difficult to obtain.

Sometimes industries find that ordinary good housekeeping and conservation of raw materials will materially reduce the quantities of waste materials

Table 5-2
Cost of Various Chains of Processes

	Est. % removal				Treatment cost ¢/1000 gals.		
BOD	COD	Phos.	Nitr.	Process Number	1 mgd	10 mgd	100 mgd
35	—	10	0	1	13.7	7.7	4.4
88	—	25	0	2,9	25.4	14.3	8.6
95	—	35	0	2,3,9	26.8	15.4	9.5
97	—	92	0	2,4,9	42.5	21.2	12.3
97	—	92	0	2,5,9	46.3	22.7	13.6
97	—	92	85	2,4,6,9	49.5	25.2	15.3
—	98	95	85	2,4,6,7,8,9	88.6	39.0	23.9
—	98	98	85	2,5,6,7,8,9	92.4	40.5	25.2

discharged into the stream. In a very few instances, the value of substances recovered has paid for all costs of treatment. However, in general, major improvements of industrial effluent quality, beyond the stage of applying standard processes, are costly.

Some forms of treatment of industrial wastes are relatively simple and cleancut. Their adoption or nonadoption is simply a matter of cost, and their effects are well known, as follows:

Heat can be dissipated by cooling lagoons, spray ponds, or cooling towers, prior to return of the effluent to the stream.

Particulate matter can be skimmed off or settled out, with the aid of coagulation and filtering, and disposed of as sludge.

Segregation of certain wastes for special handling is often an essential feature of an effective system.

Acidity can be neutralized by lime, and resulting precipitates settled out with the aid of coagulation and filtering, and disposed of as sludge.

Organic matter in industrial wastes can often be treated by aeration and biological methods, similar to the activated sludge treatments in sewerage plants.

Each major plant's situation must be carefully studied on an individual basis. In some cases, municipal sewage may be added to industrial wastes to improve the mix from the point of view of providing nutrients for the bacterial activity which constitutes the biological treatment process.

Whenever systematic studies and estimates have been accomplished, it has been found that industrial treatment processes (like municipal treatment) have rapidly increasing costs as the degree of removal of waste becomes more complete. For example, in the petroleum industry, researchers from Resources for the Future have found the following marginal costs for removing BOD from the wastewater of petroleum refineries.[4]

Table 5-3
Costs of Phosphate Removal

Plant Capacity mgd	Capital Cost $1000	Operating Costs ¢/M gals.	Capital Cost $1000	Operating Cost ¢/M gals.
1	380	29	21	4[a]
10	1600	16	72	4
100	6800	10	697	4

Note: Capital recovery costs:

30 years and 8% = .08883

25 years and 8% = .09368

[a]Given as 3.6 or 4.2¢ depending upon the process used.

70% reduction	6¢
85% reduction	17¢
95% reduction	24¢

Similarly, marginal costs of BOD removal in the beet sugar industry have been shown to increase equally sharply.[5]

Data such as these can be used to investigate whether or not certain degrees of treatment are economical. There certainly is no point upon a broad spectrum of possibilities at which a professional determination can be made that treatment up to that given point is "best practicable" or "best available." Such determinations can only be arbitrary, no matter whether made by administrators or in the judicial process.

Industrial wastes which are poisonous or otherwise highly objectionable, or which are expensive to treat, may be disposed of by pumping them through deep wells into underground formations which cannot contaminate water supply, or by trunk sewers, which carry the wastewaters to some central point for treatment in a large plant, or by concentrating the wastes for storage or disposal in solid form. In the Ruhr Valley of Germany, especially intractable and toxic wastes are segregated and transported to a special detoxication plant, where they are treated so as to render them as harmless as possible, and the residues are buried in a geologically isolated valley. Such an arrangement is carried on as an activity by the river basin authority, the Ruhrverband, described in Chapter 7.

The reduction or elimination of industrial wastes can often be done best through modification of industrial processes rather than treatment of wastewaters. This hypothesis was first advanced in the United States by Allen Kneese, based upon experience in Germany and has been conclusively proved to be the case in the U.S. beet sugar industry. Most industries using water ordinarily pay little attention to the impact of technology change in water use, but firms subject to a waste disposal charge have become interested in waste process modification and water recycling. New waste treatment requirements have brought the necessity for pollution control closer to industry's attention, and there is no doubt that changes in process modification will be effected where incentives are provided.

There are, of course, many cases where processed industrial effluent is used for beneficial purposes. Probably the best known case is the Bethlehem Steel Company's Sparrows Point plant near Baltimore. The Company, on the advice of Dr. Abel Wolman, has for years used a large amount of effluent from the Back River Sewage Treatment Works for cooling processes in the mills.[6] The Middlesex County Sewage Authority of New Jersey is starting construction of a large new waste treatment plant which should produce a high quality effluent. Negotiations with various industries are underway to find a suitable use for the resulting additional supply of water.

Other Processes

There are other types of advanced waste treatment which have possibilities. The "UNOX" process uses pure oxygen instead of air in a biological treatment process. Some large applications of this process are under construction, including the Middlesex County Sewage Authority plant on the Raritan River. A principal advantage is that the process is very compact, so that by its use an existing plant can greatly increase its capacity within the same area.

American Cyanamid will apply a carbon-adsorption wastewater treatment process to its Bound Brook, New Jersey plant to treat 20 mgd of wastewater effluent from the present high level secondary treatment facility. Mixed-media filtration through coal, sand and gravel, will be employed to remove suspended sediment. This should produce a greatly improved effluent.

Another process is the physical-chemical plant, in which no biological treatment is attempted. Various steps are included to move the organic and other objectionable material by sedimentation and various chemical additives. One of the problems with such plants is the large quantity of slude created.

Of course, the ultimate in waste treatment for removal of mineral content is the application of desalination processes, by which virtually pure water can be created from seawater. When desalination is conducted as a supplement to a large nuclear power generating station, taking advantage of waste heat, the estimation of costs is somewhat indeterminate since the allocating of costs between the two functions can be done in various ways. Costs quoted in a National Water Commission report for desalination of seawater are shown in Table 5-4.

It may be seen that the costs quoted are roughly comparable to the costs of

Table 5-4
Costs for Desalination of Sea Water

Plants	Date	Size mgd	Water Cost/M gals.
Buckeye, Arizona	1962	.65	$.69
Key West, Florida	1966	2.6	.94
St. Thomas, V.I.	1967	2.5	.90
Rosarito Beach, Mexico	1969	7.5	.85

very thorough advanced waste treatment, in plants of comparable size. However, they are much greater than the costs of obtaining additional supplies of water by reservoir storage in most parts of the country. It should be noted that desalination of brackish water costs much less than of seawater. Although desalination of seawater is currently economical only for island situations, such as Key West, Florida, or extremely arid regions, the desalination of many mineralized (brackish) waters of the Western United States may become feasible in the not too distant future.

For small communities, the usual activated sludge waste treatment plant may be relatively expensive, and difficult to operate with necessary technical efficiency. For such communities, where space is no problem, oxidation ponds or aerated lagoons have been advocated, sometimes followed by irrigation with the effluent. These have mostly been used in the West and South. In one community in New Jersey, a plan has been drawn up for such an installation. This method was selected for this locality because of the necessity of discharging the effluent into a river of very high quality water. In this case, the land disposal of effluent is to be used to avoid the deterioration of water quality in the receiving stream.

We have not even approached the ultimate in waste treatment technology. For example, an innovative waste treatment system for the Dow Chemical Company's plant at Freeport, Texas, uses a two-stage biological treatment process.[7] The conventional first stage produces an effluent high in nutrients and bacterial mass. The second process uses multitudes of tiny brine shrimp which obtain food from the bacterial masses; they and their eggs then provide a commercially salable component of commercial fish food. The use of this second biological stage avoids the costs of providing conventional sludge disposal facilities, besides furnishing a supplemental income. Other applications of aquaculture in connection with waste disposal are being devised.

Disinfection of Effluents

Until recently the EPA has required chlorination of all municipal wastewaters, and has advocated the chlorination of effluent from any treatment of urban storm water. Now, however, the states have been left free to decide whether chlorination or some other method of disinfection may be best. The most practicable alternative to chlorination may be treatment by ozone, which is used to a considerable extent in Europe, but is hardly known here.

This change in policy is due to the discovery of carcinogenic pollutants in drinking water for New Orleans (drawn from the Mississippi River) which, it is surmised, may have been formed from chlorination of effluents in some of the many cities upstream. It has been known for some time that chlorination of wastewaters or effluents containing petroleum can form chlorinated hydrocarbons, and that some of this class of compounds are among the most toxic substances known. Since petroleum in concentrations of two mg/l and above has now been shown to occur in runoff from fairly clean urban areas[8] and petroleum also occurs in municipal sewage effluents (research to measure it is now underway), there was sound reason to question unlimited use of chlorination even before the publicity about the New Orleans water supply.

Disinfection of wastes is a subject in which totally insufficient research has been done. The case for disinfection rests upon high coliform counts, presumably representing pathogenic bacteria, which it is felt should be reduced. However, as brought out in Chapter 2, the British searched systematically for epidemiological evidence of health hazards from polluted water contact, and, finding none, have abolished the use of coliform counts as a criterion of water quality for bathing. Limited but still concrete experience in the United States indicates that high bacterial and viral counts in water used for body contact recreation should not be disregarded. However, it appears that in clean seawater, sewage bacteria die out with great rapidity. As far as water supply alone is concerned, it might be better not to chlorinate effluents coming into the watersheds. The chlorination of water with considerable organic content takes large amounts of chlorine, and may create toxic or odoriferous compounds and possibly even carcinogenic substances. The raw water from a stream is always filtered and clarified in a water treatment plant before use. A stage of carbon absorption can be added if desired. The treated water can be disinfected with very little chlorine, with little risk of creating harmful substances.

The treatment with ozone, although expensive, is presumably satisfactory, based upon European experience, but it has not been given a really searching examination in the United States to determine its environmental consequences.

It may be readily understood that the subject of disinfection of effluents, including the desirability of disinfecting urban runoff, requires a lot more study. It also appears to be another subject which will not be best covered by a single uniform rule from Washington.

Sludge and Land Disposal

Even a generation ago, disposal of sludge from waste treatment plants did not seem to be much of a problem. There seemed to be plenty of open space for disposal, and farmers were often glad to have the material spread on their fields for its fertilizer content. Now, however, the matter of solid waste disposal has become much more difficult. Recent experience indicates a great deal of resistance by private landowners to accepting sludges on land, or even moving sludge across local municipalities or counties. Incineration is frequently used to avoid these problems, but air pollution may result. Cities in crowded regions such as New York and Philadelphia have tremendous problems in deciding how to dispose of their sludge and other solid wastes.

Meanwhile, in the wake of the main environmental movement there arose a strong and at times emotional advocacy to dispose of wastes by spreading them directly on land, or alternatively of giving wastewaters a preliminary treatment and then spreading either the effluent or the sludge on land. (Raw sewage was traditionally used on agricultural land in China, and sewage in Munich was in past times fed to carp which were then sold on the market, but no one in the United States has recommended such hazardous practices.) Such disposal has, of course, the environmental advantage that residual organic material and nutrients are returned to the soil, thus conserving these resources for future use.

Irrigation of field crops, forests and pasture with effluents might be beneficial. In some parts of the world, especially Australia, disposal of wastes in this manner has been carried out for years without apparent ill effect. The Corps of Engineers, in particular, was exposed for awhile to great pressure to incorporate into its water quality planning basic provisions for land disposal.

Now, however, a more balanced viewpoint is generally taken. It is commonly accepted that land disposal is an alternative worth considering seriously, but it is felt that large-scale applications should be considered very carefully. Besides the public objections which would be sure to arise in nearby communities, and the large amounts of land required, there may be dangers of insect infestations, pollution of groundwater, and buildup of heavy metals, salts, and possibly other toxics in the soil. Each of these aspects requires careful investigation. Some practical demonstration projects have been initiated, particularly in Illinois, and careful monitoring of them should give us much more information in a few years. In Santee County, California, a water district has passed effluent from an activated sludge treatment plant through an oxidation pond and from there to spreading basins. The effluent passes through natural aquifers and into a system of lakes, which are used for recreational, industrial and agricultural purposes. Spraying of effluent on strip mining wastes or forests may be advantageous. Carefully planned systems may provide real advantages without creating any environmental hazard, but the careful planning is essential.

The Interstate Sanitation Commission, with support from the EPA, contracted for a comprehensive study of the alternatives to ocean disposal of sludge from waste treatment in the New York area.[9] The ocean disposal is known to be causing environmental damage, and the EPA is determined to bring the practice of ocean dumping to an end. The study considered various alternatives, including the usual methods of land disposal and of incineration of sludges. It was decided that a limited quantity of the sludge, containing minimal amounts of heavy metals, could be disposed of by land fill, but that the rest would have the least total environmental impacts if disposed of by dewatering with filter presses, followed by pyrolysis. Pyrolysis is a heating process which decomposes the sludge in absence of air and produces gaseous, liquid and solid residues, some of which may be used as fuels. Such a system, it is estimated, could be operational by the year 1985, and the facilities would cost 400 to 500 million dollars. Based upon the institutional, technical, and legal novelty of the recommended solution, and the costs involved, it seems doubtful that such a system actually will be constructed by this date. However, what is sure is that the problem of sludge disposal is a major one, which is likely to plague metropolitan areas for many years to come.

Monitoring

The monitoring and surveillance of waste treatment systems are generally very loose. Some states visit treatment plants monthly, or even every three months, during business hours. Such schedules are totally inadequate to maintain control. Minor industrial and commercial waste-producers have carefully developed routines of discharging the bad wastes (or blowing down the boilers or discharging polluted cooling water residuals) at night, or when it rains, in order to avoid detection. Some facilities, besides their official discharge points, have some unofficial ones. To correct abuses, one suggestion has been made that locked automatic samplers should be installed on the effluent discharge lines of major companies, which would sample the effluent at frequent intervals, keeping in (refrigerated) storage the samples of the last twenty-four hours. Under this system, an inspector coming at times of a suspected spill, or otherwise unexpectedly, would have available samples from the previous twenty-four hours.

In the Ryemeads plant in England, which is an impressive, highly modern regional waste treatment plant, daily samples are taken of all of the industrial plants upstream. Each sample is tested in the basin authority laboratories to determine its toxicity towards the nitrification bacteria which perform an essential function in the treatment process. If results indicate toxicity, the offending industry is summarily ordered to close down, even though the exact cause has not been determined.

Of course, there are possibilities of complete automation of samplers, with sensors which would automatically signal a violation of effluent standards, but with our present hardware the range of pollutants which can be measured by fully automated means is too limited, and the cost would be exorbitant.

Problems

The treatment technology now being put into effect is greatly improved but still inadequate. There are still some problems with removal of nutrients, of bacterial and viral pathogens, and of heavy metals, at reasonable costs. The control of suspended sediments has not been properly related to the content of pollutants such as petroleum and heavy metals, which ordinarily are contained in or adsorbed on such particles. When detention or other processes remove certain proportionate parts of the sediment, will it retain corresponding proportions of these pollutants? Or are some of the pollutants closely allied to extremely fine particles which may not be removed? The general preoccupation with legal criteria of control has left some important technical questions unanswered.

The problem of aesthetics in treatment plant layout and design should be given more attention. It would be ridiculous, for example, to try to camouflage any but the smallest waste treatment plants as something else. However, as discussed in Chapter 7, German planners have found simple methods of layout and landscaping, which allow the incorporation of a treatment plant into a residential community without giving visual offense.

The last problems concern the economics of planning regional waste treatment plant systems. There are, of course, major economies in constructing and operating large waste treatment plants, and technical control of a few large plants is apt to be more effective than of many small ones. Based upon these facts, many analysts and administrators decided that the future lay with regionalization. It was expected that there would be major economies in replacing groups of small treatment plants with single large plants, and providing trunk sewers to collect and bring in the wastes. However, more detailed studies indicate that this may not be the case. The cost of the sewage collection systems to bring all of the wastes to a central point tends to counterbalance the savings in treatment plant costs. There are also some environmental disadvantages of the regional systems, which at first were largely overlooked, which are discussed in Chapter 8.

6

Principles of Water Quality Planning

Political Context

The basic political decision regarding water quality has already been made, namely, it will be controlled by government, with the leadership exercised at the federal level, in spite of language in the various federal acts which assert primary state responsibility for water pollution control. The questions remaining concern the manner and the institutions through which the policies of the federal government will be implemented, and particularly the type of planning which will be carried on.

The dictionary definitions of the words plan and planning (of which the broadest refers to preparation of a scheme of action) give little idea of the significance of the word in modern society. To political conservatives, governmental planning may be viewed primarily as an attribute of socialism, to be minimized in a capitalistic society; while to political liberals it is usually viewed as the process by which the interests of society are advanced against powerful and rapacious business and private interests. While political aspects are inseparable from any public planning process, the objective of this book is to establish principles and technological criteria which, when applied, will indicate objectively the relationships involved in any particular case. This should eliminate confusion, and, hopefully, restrict the role of politics in decision making to the indication of basic priorities and objectives and the final selection between technically sound alternatives.

The main exception to this present federal control of water pollution matters is in land use planning. Governmental powers of land use control are exercised primarily by local governments under authority of the states, although some limited programs are federally supervised, and there is no federal legislation providing for land use planning and control specifically in aid of water pollution objectives. This remains an active political question. However, it is possible for local governments to institute land use controls shown by water quality planning to be desirable, without any expansion of present federal powers. The political aspects become important mainly when water quality plans are devised in such a way as to control land developments.

This chapter examines the principles of water resource planning, as applied to water quality, under the assumption that the federal government, with some state participation, will control the quality of our waters by one means or another; either by means of a formal planning process, or by legislation and

123

regulations, implemented at bureaucratic discretion. However, no matter what channels and approaches are used, the principles involved in wise use of available resources are the same.

Background

In 1936, spurred by the political winds of change then sweeping the nation, the Congress passed a flood control act which established the first general criterion for initiating federal public works. This criterion was that the benefits "to whomsoever they may accrue" should exceed the costs. Although this might appear to be a mere truism, this formal statement of 1936 actually was a vast step forward from the original "pork barrel" approach. For over thirty years thereafter, the calculation of project benefits and of project costs by federal agencies was relied upon to indicate a rough order of merit of various proposals, or at least to distinguish those eligible for selection by Congress from others which Congress considered ineligible for authorization. This system was, on the whole, useful; however, it still allowed some very poor projects to be authorized. Some of these poor projects were duly built; but others, such as the project to extend river navigation across the dry plains of Oklahoma to Oklahoma City, remain unquietly in a sort of budget limbo, with interested groups periodically trying to haul them out and refurbish them.

In practice, there were two main defects with the benefit-cost approach. In the first place, the officials of the Corps of Engineers and other construction agencies sometimes simply made mistakes, or yielded to political pressure. However, a more fundamental difficulty lay with what were called intangible effects of the improvement. It was clear enough that project benefits such as flood damage prevention or provision of irrigation water could be evaluated in money terms and included in a benefit cost ratio. These were designated as tangible benefits. The intangibles included such matters as the effects of flood protection upon flood plain occupancy, or effects of reservoir construction upon fish, wildlife, and natural ecosystems. In theory, all of these intangible aspects should have been considered prior to authorization, but in practice they seldom were adequately investigated, or the congressional committees did not always give them sufficient weight. When the environmental movement gained ascendancy in the late sixties, such approaches could no longer be tolerated.

The United States Water Resources Council (the top level federal policy-coordinating body for water resources) attempted to cope with the growing environmental pressures by a new approach.[1] This was the establishment of a methodology called multiple-objective planning, under which there are two main criteria, economic efficiency and environmental quality. Two separate systems of analysis are prescribed, both of which are to be used. The planning for the economic efficiency objective quite closely resembles the comparison of tangible costs and benefits under the older approach. The planning for environmental quality involves consideration of a large class of what were previously designated as intangible benefits and costs. To unify the two approaches, a format of summation is prescribed, which attempts to insure that the two objectives of economic efficiency and environmental quality are given equal consideration. In principle, this new system is not really much different from the old one, but it does go a long way to insure full consideration of environmental aspects. However, the problem of quantifying the major environmental benefits, especially those relating to ecology, is as difficult as ever. The Water Resources Council approach provides no satisfactory general intellectual linkage between the two objective functions (i.e., the environmental and the economic). Planners and systems engineers have been forced to devise various expedient approaches, some of which are discussed in Chapter 2. Unfortunately, it is still possible to do bad planning under the new system in the same circumstances as under the old one: namely, when there is staff ineptitude or undue yielding to political influence. Moreover, an unfortunate result of the new approach is that the formal planning process has been slowed up, complicated, and made much more expensive; consequently there is relatively little new project planning through this process, either good or bad.

Further improvement in the multiple-objective planning methodology itself would do little for water quality planning. Multiple-objective criteria are not readily applicable to water quality planning, and the EPA has had minimal participation in regional comprehensive basin plans. Although in the one case of the Delaware River there was an initial attempt to evaluate costs and benefits of water quality, and a preliminary evaluation was published in 1966 by the predecessor of the present EPA, there was no further federal attempt to evaluate benefits of improved water quality in monetary terms, and little effort to evaluate them in any terms at all. EPA cooperation in "comprehensive" regional water resource planning remains quite nominal. PL 92-500 in effect *assumes* the benefits of all waste treatment proposals as being sufficient to justify achieving the stated goals nationwide. Therefore, under present circumstances, the Water Resource Council's procedures of multiple-objective planning are of limited value for comprehensive water resource investigations, and are almost totally inapplicable to planning for water quality control.

With the fall into disuse of the formally stated criteria for planning, a tendency developed to make more decisions in accordance with manifested public concern. Priorities for research and for budgeting tried to follow the headlines, a prime example being the great enthusiasm for removing phosphorus from detergents, which was to have been required nationwide by the federal government and was required by a number of states. In most cases, the removal of phosphates from detergents does little to limit eutrophication. There would still be enough phosphorus from other sources to require tertiary treatment of sewage to reduce the phosphorus below the threshold values for algal blooms;

and, if tertiary treatment is to be required, the phosphorus from detergents could be removed as well. If remaining phosphorus exceeds the threshold concentration values, it is not clear that the part removed has major value in limiting eutrophication. Therefore, despite the initial public outcry, the proposed nationwide removal of phosphorus from detergents was finally seen to be a mistake, especially as the substitute detergents also were suspected of having potentially bad environmental effects.

Most planning for water quality control is now carried out under provisions of the 1972 Act. It is done under EPA supervision, primarily by states and their subdivisions. The Act calls for basinwide planning (Section 303e), areawide planning (Section 208) and facilities planning (Section 201). The EPA has chosen to carry out the planning in a somewhat illogical order. To get work under way rapidly, major emphasis and funding was put first on preparing facilities plans. States were given grants for basinwide planning, but the sums allotted and trained personnel available were totally inadequate to do the job. Moreover, in the case of interstate rivers, no adequate arrangements were made; for example, no serious attempts were made to have basinwide plans prepared for the Delaware and Hudson Rivers. Finally, in 1975, a major start was made in areawide waste treatment planning under Section 208, and by the end of 1975 $200 million had been allotted for this purpose.[2] It is important to review these EPA planning programs and also to review some of the basic principles involved, as well as various other planning activities, which affect water quality control or interfere with it.

Evaluation of Water Quality Benefits and Costs

One of the main reasons the Water Resources Council's approaches to water quality control is hampered is that total benefits and costs of water quality improvement are virtually impossible to estimate. Some benefits can be evaluated by normal economic approaches based upon market prices, such as the value of commercial fishing, which would be harmed by water pollution. Another class of environmental effect, the recreational value of (clean) water, can be evaluated by indirect means, which in effect estimates public willingness to pay for recreation facilities. Although expensive to apply and far from rigorous by economic standards, such approaches can assign reasonable money values to recreation benefits using federal guidelines which provide approximate figures, based upon estimates of visitor days anticipated in various categories of water-based recreation. Estimates have been made of damage to water appliances by use of water containing excessive dissolved solids.[3] Less rigorous estimates have been made of damages from household use of water exceeding prescribed limits of various constituents, which may adversely affect health, stain clothes and require excessive use of soap and water softeners.[4] Methods have been devised for estimating the extra value of homesites if located in or near attractive bodies of water, and the reduction in homesite value if these waterbodies are foul and odoriferous.

In view of the progress which has been made, there seems to be no reason why a reasonably good approximation of the economic value of water quality in streams could not ultimately be developed. This would still leave out of the accounting certain intangibles, such as the value of preserving rare biological species, and some social values to society which do not appear to be susceptible to economic evaluation at all, as explained in Chapter 2. Moreover, it would be very vague regarding public health benefits, as there is little basis to evaluate the more subtle effects of various pollutants upon the health of humans or other species. Even with these reservations, it would be well worthwhile to develop benefit evaluation approaches further, as they would provide much more insight into the effects of alternative choices open to society. However, the EPA has not devoted any appreciable part of its vast resources to such matters; and, in spite of some good beginnings, little progress has been made in developing an operationally useful methodology for evaluating water quality benefits. In fact, of the total expenditures of the EPA for research through FY 1973, only two percent was devoted to research related primarily to planning approaches.[5] Friends of EPA feel that this results from Congressional priorities for immediate action expressed in the 1972 Act. In any event, regardless of what might be done or should be done, it is beyond question that we do not have now, and will not soon have, such a methodology.

The lack of real evaluation of water quality benefits precludes the use of some planning approaches which otherwise would have value. Complex systems of optimization based upon assumed measurements of benefits of water quality are nothing more than intellectual exercises. This is also a major difficulty with Dr. Allen Kneese's brilliant exposition of a system of effluent charges as a means of water quality control.[6] Although he conclusively shows the theoretical advantages of such a system, he does so by means of referring to economic benefits of water quality, which, while they undoubtedly exist, cannot be measured. Therefore, lacking measurement of the benefits, the application of the system to the real world at this time could only be on some empirical basis.

The "Principles" of the Water Resources Council suggest that where money evaluation of environmental benefits is impracticable, other quantitative measures can be applied.[7] This can be done by means of planning units such as number of visitor days, acreage of lakes, numbers of campsites, number of miles of wild river, etc. as explained in Chapter 2. However, there is nothing to assure that weights subjectively determined in this manner by any one observer will correspond to the views and judgments of any other observer.

As a closely related approach, the concept of recreation-carrying capacity is being developed by agencies responsible for river recreation and especially scenic waterway responsibilities. The term was originally applied in wildlife management to designate the numbers of animals which could satisfactorily occupy a given habitat. In this usage, since the criterion was survival, the term could be applied with a certain degree of precision. However, when applied to estimating the numbers of humans who can satisfactorily enjoy some piece of terrain such as a river, it is obvious that subjective value judgments are involved.[8] The approach is a useful one, but statements of recreation-carrying capacity should be recognized as estimates, the number arrived at depending largely upon the standard of recreational elbowroom which is assumed to be necessary.

It may be obvious in principle, but it is hard to get people to remember that costs to all segments of society should be assessed equally. There is an almost irresistible tendency on the part of the public and both local and state officials to consider that federal funds are "free," since they do not have to be raised locally. Some federal officials and segments of the public seem to consider costs incurred by industry, particularly large industry, to be of relatively inconsiderable importance. A few even appear to welcome the prospects that heavy costs will be caused to some industry by a proposed action, forgetting that in the long run, such costs will inevitably be passed on to the consumer. Water quality planning should not be diverted into efforts to alter the basic economic and social structure of our society, or to try to obtain a selfish advantage for any group or interest, whether bureaucratic, industrial, or merely personal. The only justifiable position for professional planners is that we have limited resources of materials, energy and manpower to deploy in the safeguard of our water quality, and that the means adopted should minimize the total costs to society. The 1935 mandate of Congress to consider benefits "to whomsoever they may apply" is still a good general guide. If any extra sociological advantages can be gained by water resources programs, this should be done, but general jockeying for political advantage and the advancement of political aims of special groups is highly disruptive to a planning process.

Resource Economics Investment Criteria

Although the benefit cost approach was the original method for justifying water resource projects, it is a practical method rather than a fundamental concept. The basic principle is that in a world where resources are limited and needs are great, the more valuable and productive approaches open to us should be used first. If a series of investment opportunities are ranged in order of unit productivity, starting with the most productive, and the needs for the resources produced are ranged in order, starting at the greatest need, Figure 6-1 shows quite clearly that there will be a point A beyond which no further investment will be socially productive, since the next increment would consume more resources than it would produce. Also, there may be some point B beyond which investment cannot go, either temporarily or permanently, because of constraints,

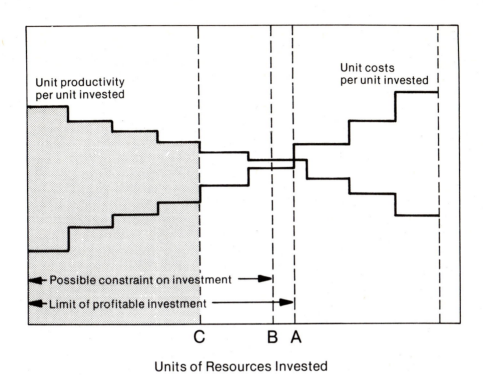

Figure 6-1. Resource Economics Investment Criterion

including inability to mobilize the resources for any reason. Of course all sorts of theoretical difficulties may remain in measuring the units of resources involved, particularly those needed to estimate unit productivity, but the principle is clear and incontrovertible.

The location of point *A* on the graph cannot, of course, be determined without complete evaluation of water quality benefits. However, we can accomplish some very important optimizations without detailed knowledge of the benefits curve. For any given or proposed level of investment, we can evaluate the alternatives which will produce the highest water quality and decide which are most desirable. Also, for any given water quality objective we can identify the lowest cost combination of improvements which will achieve it, even though the corresponding benefits cannot be evaluated, remembering that environmental and social as well as money costs are involved. Of course, this process of cost-effective planning is much more complex to carry out than to enunciate, as will be illustrated in sections which follow.

Students of economics will notice the similarity of his approach to the traditional marginal analysis of supply and demand. There are serious qualifications to use of the classical supply and demand analysis on account of the many imperfections in the hypothetically free market; but these qualifications do not apply to the marginal resource investment criterion, since it does not represent market phenomena.

Program Interrelationships

General approaches to management and planning have much to recommend them, since almost all aspects of our lives relate to many other aspects. The integration of various types of specialized approaches into overall planning is politically alluring, but knowledge and expertise can be submerged in the process, and there is danger of losing contact with reality. Water quality planning requires detailed knowledge and expertise, however, it should not be conducted in isolation from other related matters. The most important of these interfaces include water supply, land use planning, erosion control, flood control, environmental impact analysis and residuals management, all of which are discussed in sections which follow.

Water Supply and Water Quality Relationships

The interface of pollution control with water supply planning is of major importance. There is a real need for a better approach. Although the EPA now sets standards for drinking water under the Safe Drinking Water Act, it has no responsibility for developing water supplies, and even the incorporation of low flow augmentation provisions into water quality planning is not allowed. What is probably even more significant is the failure to relate water quality planning even in principle to possible reductions in the quantities of water flowing into the streams. The rationale for this, although incorrect, seems logical. Since the 1972 Act states as a goal that all pollution of our streams should be eliminated by 1985, it might be assumed that the streams after that date will be clean and unpolluted regardless of the quantities of water flowing in them. If one takes the law at face value, the conclusion seems plausible. However, the nonpoint sources, illegal dumpings and urban runoff will be sufficient to pollute our streams to such a considerable extent, and the costs of waste water treatment are so enormous, that by 1985 many of our streams will still fail grossly to meet existing water quality standards, even if the flows in them are maintained unimpaired. Meantime, however, regional, state, local, and private plans are being made to develop water supplies further, based largely on the concept of using the maximum safe yield, (use of the maximum quantity of water available based upon flows during a prolonged drought) so that there is no real assurance that prevailing low flows in the rivers will not be greatly reduced.

This is a problem which must be faced. It would be quite meaningless to rely for policy on the slogan "dilution is no solution for pollution." The fact is that, as shown for the Willamette River, a large river flow can handle satisfactorily a larger pollutant loading than a smaller river flow. The separation of water quality planning from water supply has allowed both programs to ignore a key relationship affecting the objectives they hope to achieve. The state and federal fish and wildlife agencies do what they can to encourage maintenance of low flows, but such action is not enough. The protection of needed low flows in our streams should be an integral part of basic processes of water quality planning.

Water supply is also affected by water pollution in another way: surface runoff or waste discharges enter the aquifers from which groundwater is drawn. In some cases, groundwater may be affected by sewage effluents or sludge disposed of on the ground, or by septic tanks. Many watersheds in metropolitan areas are rapidly becoming developed for housing, towns, and commercial and industrial facilities of all kinds. Because of nonpoint as well as point sources of pollution, these developments are adversely affecting the groundwater to such an extent that water from these watersheds may become unsafe for human consumption even after treatment. In such cases, a difficult choice must be faced sooner or later; either control land use, or abandon the use of the watershed for supply of potable water.

One water supply aspect in which the EPA acts vigorously is that of encouraging recycling and reuse of water. The possibilities of industrial or agricultural reuse of water are very promising, and even reuse for domestic purposes may some day be acceptable.

Erosion and Sediment Interface

In the interest of water quality, it is necessary to reduce the high quantities of sediment reaching our waterways, which are quite harmful to the ecosystem and physically clog the channels. Control of sediment from agricultural and forest lands has been an objective of the Soil Conservation Service for a long time (with considerable success), but there is far less control of sediment from construction sites and urban and industrial areas. Some states are now moving to exercise control over sedimentation from such sources.

Insufficient attention has been given to environmental and physical damage done by sediment in water, and to sources of sediment, as discussed in Chapter 4. Sediment from major point sources will soon be controlled reasonably well, but very little data exists as to the extensive sediment output from streets and other urban runoff, and from the many unrecorded commercial and industrial sources. Sedimentation and erosion aspects such as these are widely neglected in most metropolitan area planning.

Flood Control Interface

In many cases flood control and water quality control should be linked through water storage or detention plans. The benefits of reservoir storage to water quality are apparent where the question has been seriously studied, as described later in this chapter for the Willamette River. There is another linkage, however, which is potentially much more frequent, and this is the use of urban runoff detention to the joint benefit of flood damage prevention and water quality. The retained water may be given primary treatment directly, or simply held until passed through the treatment plant at a manageable rate.

Environmental Impact Interface

Finally, it should be apparent that water quality planning and environmental impact analysis are closely linked. An environmental impact analysis typically requires a responsible agency or company to examine the environmental consequences of some proposal to build, reconstruct or change the operation of some facility. None of the environmental impact methodologies currently in use are really satisfactory, although they do provide frameworks which are conceptually useful to the extent that the necessary evaluations can be provided. Recent editorials in *Science* provide strong criticism and an official rebuttal defending environmental impact statements.[9,10,11] However, even the rebuttal admits that "many EIS's are too long, they often include extraneous material that is neither analytical and predictive, and the scientific quality of most is below desired standards." It may be added that the environmental analyses prepared by the nonfederal agencies and private companies are apt to be of a lower quality than the federal impact studies.

Where an effect of the proposal being evaluated is to change loadings of pollution in the stream, a definitive analysis of conditions affected may involve substantially the same data gathering and technical considerations as that involved in studying various pollution control alternatives. Of course, many environmental impact analyses may involve quite different matters, but the interrelationships with water quality planning are often substantial. It seems that there is inherently considerable duplication involved between environmental impact analysis and planning. Some consideration should be given to carrying out water quality planning in such a way as to facilitate environmental impact analysis. This matter is discussed in more detail in Chapter 8.

Corps of Engineers Planning Approaches

Curiously enough, the EPA is not the only agency responsible for comprehensive water quality planning under federal law; the U.S. Army Corps of Engineers is also active in the field. Several years ago Congress, presumably impatient at the extreme slowness of the EPA to carry out general area planning, and possibly also tired of the Corps' preoccupation with dam building, authorized the Corps of Engineers to carry out a series of water quality control studies of metropolitan areas.[12] A major aspect of these studies was the consideration of alternative methods of handling wastes, including land disposal. Only two of these studies were completed in final form by 1975, but thirty-two more are underway at an average cost of $1.5 million each.[13] This represents a major effort. The Corps of Engineers suffers a major handicap in that it does not have general statutory authority for implementing water pollution control as such (although it does have a responsibility under the 1972 Act to control fills and dredging), but it has advantages in that it does have national responsibilities for flood control and river navigation and substantial interest in water supply, besides a considerable general engineering expertise. This pollution control planning by the Corps of Engineers is an anomaly that would not have occurred if the EPA had exercised its planning function more vigorously.

Unlike the EPA, which usually requires planning to be conducted by states and state subdivisions, the Corps of Engineers does the work itself, within the framework of the U.S. Water Resources Council's policies. What concerns us here is the methodology by which the work is carried on. The Corps of Engineers' approach to planning its normal construction projects is as shown in Figure 6-2, while its approach to water quality planning is illustrated in Figure 6-3. At first glance, these procedures appear to be entirely different; but the essential major planning approach is the same, namely, developing goals and objectives, preparing a series of alternatives to achieve the goals, and choosing the best from among them. The most fundamental difference between them is that the final report concerning construction projects is made by the Corps of Engineers to the Congress, whereas the water quality plans can only be implemented by the EPA.

It may be instructive to consider how the Corps of Engineers proceeded to plan for a given area, the lower Merrimack Basin in Massachusetts. The report in question covers an area which had a population of 468,000 in 1970 and included four cities: Lowell, Lawrence, Haverhill and Newburyport.[14] The stated objective of the study was to provide water pollution control construction alternatives, costs and schedules required to meet the target dates and criteria of the 1972 Act.

One of the principal aspects of the study was the consideration of land-oriented treatment systems involving ultimate disposal of waste residuals on land, as well as water-oriented treatment systems. The water-oriented systems included various types of advanced waste treatment to meet the successive requirements of the 1972 Act. Steps past secondary treatment included nitrification (to remove ammonia), denitrification (to remove nitrates), filtration, carbon adsorption, disinfection, aeration, and of course treatment of sludge, to include incineration. The land-oriented systems included consideration

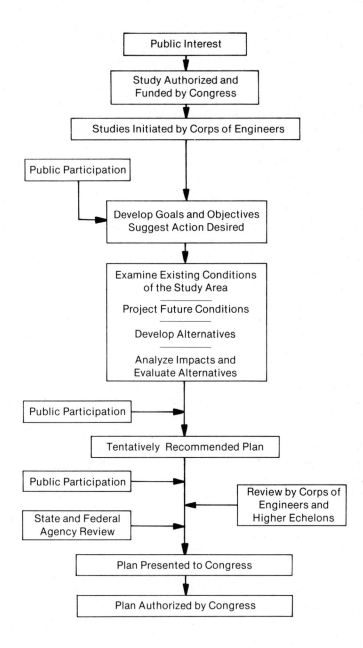

Source: U.S. Army Corps of Engineers "Merrimack Wastewater Management," Summary volume and seven appendices, New England Division, November 1974. Courtesy of the U.S. Army Corps of Engineers.

Figure 6-2. Corps of Engineers, Construction Project Planning Process

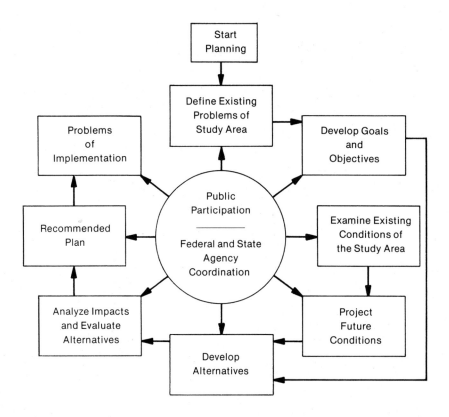

Source: U.S. Army Corps of Engineers "Merrimack Wastewater Management," Summary volume and seven appendices, New England Division, November 1974. Courtesy of the U.S. Army Corps of Engineers.

Figure 6-3. Corps of Engineers, Water Quality Planning Process

of two alternatives for land disposal of wastes: spray irrigation and rapid infiltration, the latter only being used for effluent which had already been subjected to nitrate removal, disinfection, and either filtration or storage in a lagoon, after the initial stages of secondary treatment. No mention was made of removing heavy metals or other pollutants. Both the land and water-oriented systems were designed to comply with the 1985 goals of the 1972 Act. Unit costs are illustrated in Figure 6-4. It will be noted that, although the capital costs of the advanced waste treatment are somewhat less than 2.5 times the capital cost of secondary treatment, the operation costs are about four times as much.

The principal alternatives considered in this planning were various groupings

TREATMENT PHASE

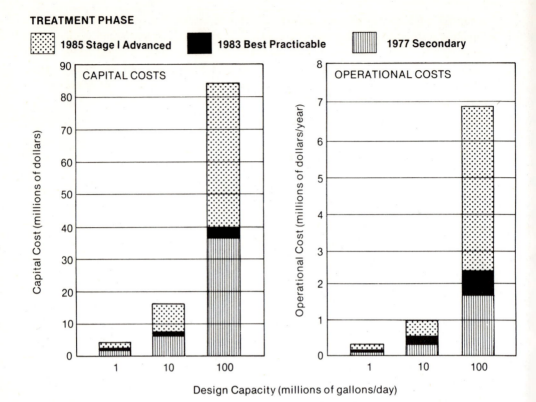

Source: U.S. Army Corps of Engineers "Merrimack Wastewater Management," Summary volume and seven appendices, New England Division, November 1974. Courtesy of the U.S. Army Corps of Engineers.

Figure 6-4. Costs of Advanced Waste Treatment Systems, Merrimack River

of waste treatment plants, ranging from a relatively decentralized system of eleven water-oriented plants to a regional system of two main treatment plants to handle the entire areas, with four other plants sending their secondary effluent to the main plants. Centralized and decentralized land treatment systems were also considered. The anticipated economic benefits of the centralized and regional systems failed to be realized, and the program of lowest capital cost was a decentralized water-oriented system.

By combining elements of various of the basic alternatives, a "least cost" plan was derived costing less than any of the original plans. A second hybrid plan was the "Impact Alternative," which minimized anticipated biological, hygienic, socio-economic and aesthetic disadvantages. A third hybrid plan balanced the potentials of idealized systems against real considerations of political viability. Finally a recommended plan was selected, which was largely decentralized, with some land features. The total capital cost would be $722 million, of which $389 million would be for collection systems, $162 million would be for secondary treatment, and $170 million for advanced waste treatment facilities. The initial yearly operation and maintenance cost for the entire system would be $13 million.

This would be an expensive facility for this region. If amortized at eight percent interest over a forty year period, the total annual costs would be $73,000,000 or $155 annually per person involved. A similar expenditure for the United States on a national basis would amount to $33 billion annually. Obviously, this is expensive treatment.

In considering the report as a whole, despite the impressive ideals expressed, the careful methodology and the tremendous detail in the sixteen weighty volumes, one is struck by the rigidity of the constraints within which the plan was conducted. Apparently, the Corps of Engineers planners felt that they had little choice. Since the plan could not be implemented unless accepted by the EPA, only the most rigid adherence to the schedules of the 1972 Act could be permitted, despite the fact that it must have been apparent to all that they were totally unrealistic.[15] There was an implicit assumption in the planning process that the zero discharge formula of 1985 is the objective towards which all planning must be focused, and all optimizations directed. Nowhere is the question asked, would some lesser degree of treatment at lesser cost provide an adequate result? In other words, the answer to this important question was assumed in advance (in compliance with the law of 1972 and supposed EPA policy), and thus the huge effort (and sixteen-volume report) were directed towards an unreal goal. The analysis does some useful work in comparing costs and disadvantages of centralized and decentralized systems of various kinds, and these comparisons dispelled the popular illusion of the economy of large regional waste control systems, at least for this study area.

However, there are a number of alternative approaches which were not adequately explored by the Corps report. The conclusion that storm sewer runoff and other nonpoint sources need not be controlled appears to have been arrived at after only a cursory analysis. There was no comparative analysis of the costs of reducing pollution from urban runoff as compared to extra costs of advanced waste treatment. There was no real investigation of the necessity for controlling nutrients in this river, so close to an estuary, although very large costs were programmed for the purpose. It might well be that nutrients coming into the river from nonpoint sources, which would be uncontrolled, would be sufficient to nullify any advantages of nutrient removal in limiting eutrophication. Also, the heavy metals were not adequately studied, although it was recognized that they have adverse biological impacts in this river. All in all, while the report has considerable substance and provides a great deal of useful data, it does not constitute a really comprehensive plan of water pollution control.

Other Planning Approaches

Wild and Scenic Rivers

The reservation of areas in a totally undeveloped condition certainly represents one form of water pollution control planning, and it is the ultimate in consideration of ecological interests. In October 1968 the Wild and Scenic Rivers Act established the principle that certain selected rivers and their immediate environments, possess outstandingly remarkable scenic, recreation, geologic, fish and wildlife, historic, cultural and other similar values, and are to be preserved in free-flowing condition for the benefit of present and future generations. Classification of such rivers is as follows:

Wild river areas	inaccessible except by trail. Vestiges of primitive America.
Scenic river areas	some roads.
Recreational river areas	have impoundments or diversions.

Both wild and scenic areas are to be kept unpolluted and free of impoundments. The designation of areas under this act is useful, but this approach cannot be carried very far in most regions without impact on economic interests.

Consulting Engineers' Approaches

Many public agencies call on consulting engineers for studies of cost effectiveness of various pollution control alternatives, some of which have been cited in previous chapters. Michael Sonnen, in a summary of such experiences, has found from many studies in the West that the cost of carefully planned waste collection and treatment alternatives fall in a range of five to twenty-dollars per person per year, with "best available treatment" being provided near the upper end of the range.[16] He finds that in a given situation centralization or regionalization of various elements does little to change the total system costs. (Compare with costs from the Merrimack Study, noted above.)

Former Federal Water Pollution Control Planning

The predecessor agencies of the present EPA had a Congressional mandate to carry on comprehensive water pollution control planning prior to 1972. Some of the "plans" produced under this authority were little more than lists of proposed sewage plant improvements appended to some bland descriptive material. However, there was at least one serious attempt to prepare a comprehensive river basin plan: the Delaware River Comprehensive Study.[17] This study was well funded, and was aided by the four states concerned, by the staff of the newly created Delaware River Basin Commission and by some highly qualified consultants. The preliminary report, issued in 1966, was a very interesting document, which is summarized below.

The report made an evaluation of fishery benefits expected in the Delaware Estuary due to various alternative programs of water quality control. Shad, sturgeon, striped bass, weakfish and white perch were at one time important commercially. Between 1885 and 1903 the commercial fishery was at its peak with annual catches amounting to twenty-five million pounds, which would be worth about $4.5 billion today. The catch continued to decline from that peak to a level of approximately eighty thousand pounds, worth only $14 thousand annually. This decline may be attributed to: (a) industrial and municipal waste discharges, (b) over-fishing, (c) introduction of predators into the area, (d) damage to spawning areas from dredging, and (e) damage to the estuary from urban and farm runoff.

The study proceeded by considering five alternative pollution control plans, the goals of each being defined in terms of twelve water quality parameters. Each of the plans referred to reaches of the river where various water uses would be made suitable from a water quality standpoint. The most expensive plan provided for a minimum DO of 6 ppm, which would permit sport and commercial fishing and the passage of anadromous fish at all times during the spring and fall periods. The total price of achieving this level of water quality was estimated to be $460 million. Other less complete degrees of control would, of course, have lesser cost, but would correspond to lower threshold concentrations of dissolved oxygen at certain months of drought years, which would in turn reduce the proportion of anadromous fish which could pass. There were very rough estimates of benefits of the plans to commercial fishing.

This report, rough as it was, was a great improvement over anything that had preceded it. After the preliminary report was issued, various meetings were held, and the Delaware River Basin Commission proceeded on this basis to adopt a dissolved oxygen standard for the estuary, and to specify organic pollution (BOD) loadings permitted to various industries and municipalities, based upon this analysis.

Curiously enough, after such a propitious beginning, the *final* report, which was for several years promised for "next year," was in the end quietly buried by the EPA, which had apparently decided that it did not want to bother going through such a burdensome planning process in other basins, when it could conduct its business more easily through other means.

Analysis of the Willamette River by the
U.S. Geological Survey

The U.S. Geological Survey for over a century has been a data-gathering agency of high professional standards which does not engage in "planning" of proposed improvements. Nonetheless, this agency has started to make analysis of water quality in various river basins, starting with the Willamette, which constitute an analytical framework upon which planning decisions can be based.

David Rickert and associates in the USGS have reported on pioneering analytical work in the Willamette study and their findings are of great interest.[18,19] The Willamette is the largest river in the United States on which all point sources of pollution receive secondary treatment; therefore it provides a unique opportunity to predict the amount of improvement likely to result from various alternatives of further treatment.

The study started with a very thorough analysis of the sources of pollution in the river, of the river's hydrological characteristics, and of water quality relationships within the river. A model for water quality analysis was derived only after the basic relationships had been thoroughly understood. The model was then used to calculate the consequences of various alternative courses of action. Some of the findings are of major policy interest.

One of the most important results of the USGS analysis was to show how important the low flow augmentation by reservoir storage (provided in the interest of navigation and hydroelectric power) has been to water quality in the river. It has been demonstrated that achievement of dissolved oxygen standards will require continuous low flow augmentation. In order to obtain the goal of forty percent DO saturation in the Willamette River, a decrease in summer flows of the river from 5500 cfs to 3500 cfs would require 1950 BOD loadings to be reduced by sixty-five percent. If the normal reduction in BOD loadings were used, the reduction in flow would reduce the DO level to less than half of what it would at sixty-five percent reduction. Since in many metropolitan areas the identified point sources are only half or less of the total population loading, it is apparent that major reductions in flow may render the achievement of desired water quality levels entirely impracticable, or prohibitively expensive. Such aspects must be brought into any serious comprehensive planning for water quality.

The study showed that the major cause of present oxygen depletion in the lower Willamette River is the point source loading of ammonia, mostly from one source, treatment of which will be extremely useful. It was also demonstrated that removal or partial reduction of what is an apparently heavy benthal oxygen demand would greatly improve dissolved oxygen levels in a five-mile section of the lower river, but the feasibility of doing this has not been determined. Another conclusion of interest is that during the low flow period of 1974, nonpoint sources contributed about forty-six percent of the total basinwide loading. The nonpoint loading represents mainly diffuse sources not potentially amenable to removal by pollution control programs. Lastly, the BOD loading from municipal treatment plant effluents presently exerts a relatively small impact upon DO levels. Increased efficiency of treatment might be useful, but its usefulness could best be determined after other more apparent improvements in ammonia and benthal demand have been accomplished.

It is quite apparent that this USGS analysis, uninhibited by the constraints and preconceptions of the 1972 Act, has shown how pollution in the Willamette Basin actually arises, and what approaches would be most effective to improve it. This sort of analysis is not a complete planning process, since it is not carried through to the stage of producing a plan. However, it shows how to avoid needless wasting of money on advanced waste treatment when much more effective approaches may be sufficient to meet environmental requirements. The valuable insights developed in this case illustrate the great desirability of such an analysis in any basin where water quality is an important objective of planning.

Residuals Management Interfaces

Some aspects of water pollution control are closely related to solid waste and air pollution problems. Studies of Resources for the Future have emphasized what is called a residuals management approach, which recognizes that waste materials are partially interchangeable as solids, liquids or gases. For example, in its sewage treatment plants, the City of Philadelphia removes waste constituents that would otherwise flow into the Delaware River. They are barged out to sea in the form of sludge. However, the sludge is adjudged to be harmful to the marine ecosystems, so the EPA has required the city to eliminate all such marine dumping within the next few years. This leaves the city with two alternatives, either dispose of the sludge on land, or burn it. If disposed of on land, it is a solid waste problem, but undesirable constituents may leach out into the soil, changing it back to a water problem again. If burned, the incinerator gases may pollute the atmosphere.

The officials of Alabama have pointed out a solid waste problem arising out of requirements to remove small quantities of chromium from wastewaters. The amount of chromium is said to be sublethal (only a few parts per billion), but when removed from the water it must be disposed of as solid waste in more concentrated form. In practice, it is stored in a warehouse in fifty-five gallon drums, which ultimately will be disposed of as solid waste on land, with the possibility of leaching into groundwater.

In general, the disposition of solid waste residuals from water treatment processes is creating increasing difficulties. Some years ago, it was usually possible to find farmers willing to accept any quantity of municipal sewage sludge directly and spread it on agricultural land in order to obtain the organic and other nutrient content; however, increased mechanization and greater dependence upon chemical fertilizers have made this practice less acceptable. Moreover, since many municipal systems now receive large industrial waste loads, the sludge is apt to contain considerable contents of heavy metals or other toxics. Therefore, careful consideration must be given sludge disposal on land to avoid the buildup of undesirable concentrations of heavy metals, nitrates or other pollutants. Even if burned, the residues must still be disposed of either on sea or land.

The new requirements for advanced waste treatment often create greatly increased sludge volumes. Situations such as those in Philadelphia are becoming more numerous as environmental concern closes previous avenues of disposal of unwanted materials.

The idea to combine the approaches to problems of water pollution, air pollution and solid waste disposal is appealing, and many government agencies have been organized so as to assure an integrated approach. When a single agency (such as the EPA) has responsibility for all three of the residuals management aspects, there is some assurance that major interrelationships will be coordinated. However, the technology of each of the three pollution aspects is so complex and so different, and the respective problem areas are geographically so dissimilar, that little real progress has been made in developing a unified residuals management approach. There are some important residuals interfaces, as where "sanitary" land fills constitute major sources of water pollution, but these are usually handled on an ad hoc basis.

Levels of Planning by the EPA

PL 92-500 requires water quality planning to be conducted by states and other agencies at three levels. The largest areas are to be covered by basin planning (Section 303 (e)), smaller areas within basins by areawide planning (Section 208) and works pertaining to a single municipality or vicinity are covered by facilities planning (Section 201).

It would have been logical to have conducted the basin planning as first priority, then to develop areawide plans within the basin plan, and finally to outline details in the facilities plans. However, the EPA has adopted a much different approach based on a desire to accelerate construction of waste treatment facilities. The basin planning is mainly considered to be the responsibility of the states and although small EPA grants have been made for that purpose, basin planning has generally languished. The objective of the basin planning is stated to be the implementation of the states' water quality standards, with emphasis on point source pollution control. The states' water quality standards include quality criteria depending on a classification of the waters of every basin as to future uses, such as drinking, fishing and swimming.

With those considerations, the basin plan should determine whether the EPA's nationwide effluent limitations for point sources will be sufficient to allow the receiving waters to meet the water quality criteria for their future uses. If not, the basin plan should determine "total maximum daily loads" for pollutants such that the water quality standards will be satisfied in all portions of the basin's waters; this is to be the basis for possibly more stringent effluent limitations. The plan also should include compliance schedules for the point sources and priorities for the construction of public treatment works. Such plans should be implemented through the discharge permits program of the individual state and through all Section 201 plans within the basin.

Since the stated objective of the basin planning is to determine justification for stricter waste treatment standards than provided for in the 1972 Act, which most analysts regard as already very strict, and as only very limited federal funds have been provided, there has been little urgency on the part of the states to complete such plans. Moreover, state responsibility is hardly the best approach to planning in interstate basins, or for coordination of basin planning with federal planning for flood control water supply, irrigation, etc.

The main impetus of EPA and local interest has come from the Section 201 facilities planning; federal funds have been made available for this purpose, and EPA has pushed hard to complete planning as rapidly as possible. The main objective is construction of wastewater collection and treatment facilities. Population projections and projections of industrial and commercial activity are developed, or adapted from previous regional studies. Engineers and planners then formulate the wastewater conveyance and treatment plans and cost estimates. Provided that the 201 plan is approved, the local governmental unit may then apply for federal and state grants for the design and construction of the facilities.

Normally, each facilities plan should conform with basin and areawide plans, but in practice every effort is being made to complete facilities plans before the areawide or basin plans can be made effective. Official policy formally requires ongoing facilities planning to be coordinated with ongoing areawide planning, but in the 208 Plan on which the author is engaged, firm EPA instructions have been issued that the conduct of 208 planning is "not to interfere with" the ongoing facilities planning.[20] It has been made quite clear by written instructions that the results of completed facilities planning, even if not yet implemented, are not to be changed by any findings of the 208 planning.[21] Similarly, NPDES permits which have been issued will apparently remain unchanged. Any new permits and any further facilities planning will supposedly be consistent with the 208 plan, but if the major point sources have all been covered previously by permits and/or facilities plans, it is not clear how or when the results of the Section 208 areawide plan will be applied.

The eighty-five Section 208 planning areas initially funded covered only about five percent of the nation's waterways. It was subsequently ordered by the courts, in 1975, that the process must be extended to all parts of the United States. This extended planning has been made the responsibility of the states, but funding for the additional areas remains uncertain.[22,23] It appears probable that progress for the remaining areas will be very slow, unless major additional federal funding becomes available.

Section 208 Areawide Planning

In principle, the Section 208 planning follows some basically sound approaches. The instructions emphasize that cost-effective and institutionally feasible plans are to be selected.[24] The following sequence of overall planning features is specified.

A. *Identity problems.* The pollution problems should be identified in terms of their relative impact on water quality. Similarly, existing institutional problems impeding solution of water quality problems should be identified.
B. *Identify constraints and priorities.* Both technical and management constraints should be identified. Priorities for solving water quality problems should be established.
C. *Identify possible solutions to problems.* All reasonable regulatory and management control methods should be identified.
D. *Develop alternative plans.* Alternative integrated technical and regulatory control methods for municipal and industrial wastes, stormwater control, nonpoint source control, and growth and development should be combined into areawide plans. Comparable alternative options for the management of these plans should also be identified.
E. *Analyze alternative plans.* The alternatives should be evaluated in terms of minimizing overall costs, maintaining environmental, social, and economic values, and assuring adequate management authority, financial capacity, and implementation feasibility.
F. *Selection of an areawide plan.* A specific procedure should be defined for monitoring plan effects and developing annual revisions to the plan.
G. *Periodic updating of the plan.* A specific procedure should be defined for monitoring plan effects and developing annual revisions to the plan.

The EPA has defined the cost-effectiveness analysis to be used. It should be a systematic comparison of alternatives to identify the solution which minimizes total costs to society over time to reliably meet given goals and objectives. Since Section 208 (B)(2)(E) specifies that the plan should document the economic, social, and environmental impact of plan implementation, the local economic impact (in addition to resource costs) should be included in the total costs to society. Thus the total costs to be minimized should include:

1. resource costs
2. economic costs
3. social costs
4. environmental costs

Effectiveness refers to meeting the 1983 goals of the Act while providing for the highest practical degree of technical reliability in the pollution control alternative that is chosen.

One feature which considerably detracts from the value of Section 208 planning is the requirement for specified interim outputs after six to nine months of the planning period, including the definition of service areas, planning areas and the treatment levels to be required for municipal treatment works within the designated area. These interim findings are intended for use in facilities planning. Since the Section 208 planning usually involves data gathering and complex analysis processes for nonpoint sources and urban runoff, which cannot possibly be completed within six to nine months after the grant is first made available, it is apparent that such interim determinations can only be made on the basis of the "best practicable" and "best available" guidelines of the 1972 Act, or on opinions without additional data. This provision appears to indicate a policy of the EPA that the 208 planning will be used only to place a requirement on municipalities to control nonpoint sources and urban runoff, and not to permit any reduction in waste treatment requirements, even if the Section 208 planning might indicate such a tradeoff to be environmentally effective and cost-effective.

Figure 6-5 shows in simplified form the various processes for Section 208 planning, according to general instructions. Figure 6-6 shows the flow chart for the Section 208 planning of the Middlesex County, New Jersey area. Although the form is different, the processes are essentially the same, and they amount to a sound approach. However, the compressed time schedules allowed were seriously inconsistent with technical requirements of sound planning. According to the guidelines received for Middlesex County planning, the initial low flow data gathering for modeling which was required, had to be conducted during a wet (1975) October (which does not simulate conditions during critical summer low flow periods) and a verification run had to be completed by the end of July 1976, which might or might not include a satisfactory low flow period. Vigorous efforts had to be made to obtain a dispensation to extend the period at least until the end of August 1976 to complete the data gathering, which may or may not be sufficient. To understand the suitability of the initially specified period, it must be realized that storm runoff evaluations require frequent hand measurements throughout the period of the storm. Such field work is onerous at any time of the year, but it is dangerous and impracticable during New Jersey winters. Moreover, much of the time the ground is frozen, and the slow rains and snows of winter cannot be used to simulate results of summer thunderstorms without some totally unjustified assumptions.

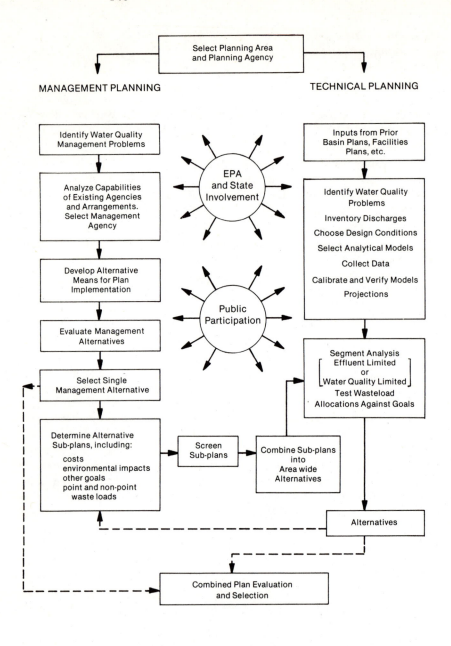

Source: U.S. Environmental Protection Agency. "Guidelines for Areawide Waste Treatment Management Planning." Washington, D.C.: Environmental Protection Agency, August 1975.

Figure 6-5. Section 208 Planning Process

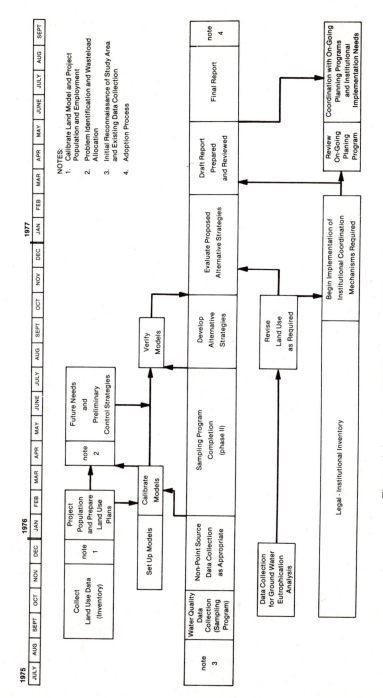

Figure 6-6. Section 208 Planning, Middlesex County, N.J.

The requirements for Section 208 planning include such prolonged formal processes of plan evaluation and coordination that there is serious danger of squeezing down the actual work of data gathering and analysis to the point of total inadequacy. It is technically unsatisfactory to do low flow modeling of critical summer periods without obtaining data on at least two low flow summer periods; one to prepare the model and the other to verify it. And it is equally necessary in urban areas to obtain data from storms in different seasons of the year, including particularly summer, before making estimates of pollution from urban runoff and nonpoint sources. The prolonged processes of plan coordination would be quite useless if the technical basis for the plan were unsound.

The guidelines for Section 208 planning include detailed instructions requiring coordination with other EPA planning, such as air quality and solid waste planning, and also with other planning efforts related to land use, including HUD, transportation and coastal zone management plans. All of this is sound and necessary, but it does not answer the need to consider possible future changes in flow regimen, either favorable or adverse, and coordination with water supply and flow augmentation planning. The only specific references to stream flow in Section 208 planning call for using severe drought conditions as the basic critical conditions, with a possible alternative, such as a summer storm condition following several weeks of dry weather.

The Section 208 planning has three major defects. In the first place, it generally bypasses both states and basinwide agencies, establishing EPA contracts with state subdivisions or local agencies. These agencies are charged with planning for areas that may include small basins in totality but often apply to a truncated portion of a large basin. For example, in the case of the Section 208 planning for Middlesex County, New Jersey, only a portion of the Raritan Basin is included in the study. Certain upstream portions of the Raritan Basin are charged mainly to another 208 planning agency, with data gathering and water quality modeling the responsibility of Middlesex County, but analysis thereafter being handled by the other agency. Large basin areas to the west have no designated 208 agency at all, but will be planned later by the state. The lower part of the Raritan River is included both in Middlesex County 208 planning and in 208 planning by New York City, and the Raritan Bay belongs to New York City exclusively, although estimation of pollution loads entering it is part of the Middlesex County Section 208 planning. The EPA staff is doing the best it can to make this awkward arrangement work, but chances for delays and confusion are obvious.

The second major defect of Section 208 planning is that Congress specified a two-year time period to accomplish it. If the job is to be done properly, this is entirely impracticable. Not only is there only a little fragmentary data on pollution from urban runoff, combined sewers, and nonpoint sources, but the technology for evaluating such pollution and incorporating it into the regular point source analysis is in process of development. As is brought out in Chapter

4, expedient methods must be devised to incorporate urban runoff pollution, which occurs primarily as a result of storms, into the usual water quality modeling that is regularly conducted at times of steady state low flows. When consideration is given to the needs of data gathering as well as modeling and analysis, to all of the prolonged processes of public participation and coordination required by the EPA's voluminous regulations, plus to the additional red tape of the other participating public bodies, a basic three-year period would be much more realistic.

In order to meet the schedule of Section 208 planning, something has to give, and there is strong pressure to make that something the data gathering and technical analysis. Administrators can clearly see the protection provided by lengthy coordination procedures, and both courts of law and administrators appear to be impressed by mathematical models, with all of their apparent certitude. Some comments on the dangers of misuse of models without data are given later in this chapter. Unfortunately, in order to meet the tight deadlines, limited budgets, and insistence on minutiae, the technical substance will be cut out of the 208 studies, particularly with respect to the determination of urban runoff and nonpoint source pollution, leaving plans supported by a facade of modeling with little substance behind it.

The third major defect in Section 208 planning is that there are no proper institutional provisions for carrying out its findings either in law or in governmental organization. The EPA guidelines require that all alternatives recommended be those that can feasibly be carried out, presumably with present laws and institutions. Even if a reasonable attitude is taken regarding alternatives, it still remains true that the 1972 Act currently provides for funding only for treatment of wastewaters, so that other alternatives, even if more economical, might not receive federal support. As long as the federal government assumes the greater part of the cost of certain programs of wastewater treatment, it is extremely unlikely that local or state agencies will proceed with any alternative at their own expense, except under very special conditions.

The other institutional difficulty is that, in most cases, no governmental organization exists to carry out comprehensive wastewater management plans, even if approved. Under the present situation, the federal EPA and the states have virtually absolute authority over point source discharges, through financing of treatment plant construction and through the system of NPDES permits described above. However, authority and procedures are unclear as regards land use controls and nonpoint sources. The local agencies performing Section 208 planning were chosen in compliance with EPA criteria; in particular, they were required to have a land use planning capability. However, they were not required to have powers to enforce land use controls, or any function or expertise whatsoever in the water quality field. In many states, land use is substantially controlled by the municipalities, and not by the intermediate-level agencies charged with Section 208 planning. The nature of Section 208 planning calls for the consideration of various structural—or physical programs—and nonstructural alternatives for controlling urban runoff and nonpoint source pollution, and if such a plan is to be put into effect and carried out, some new laws and governmental organizations will need to be created. Since these implications are not generally realized or accepted, it is likely that the plans for Section 208 will not be used, because the necessary institutions do not exist and cannot be created in time. We have good examples in France, Germany and Great Britain of the creation of new regional institutions for water planning purposes prior to specific planning (Chapter 7), while in the United States we have created only a few basin commissions whose pollution control mandates, if they existed, have been so weakened by the 1972 Act as to be hardly worth citing.

Therefore, while the Section 208 studies potentially constitute a major advance, there are serious dangers that that potential will be only partially realized, due to unrealistic time and money limitations, weakness of regional institutions involved, planning conducted on truncated portions of basins, lack of data base and technology, insufficient attention to water supply and stream flow, and lack of legal and organizational means for carrying out the completed plans. Some suggestions for improved approaches are given in Chapter 8.

Mathematical Modeling Limitations

The general lack of scientific knowledge and data concerning environmental effects of pollution has not prevented the rapid growth of mathematical modeling of water quality. The wide use of computers has immensely facilitated all computations and has also enabled the possible use of systems approaches, which encompass an entire series of analytical steps in a single computer routing or a structured series of computer routines. This is something different in kind from older analytical methods. Instead of having an individual conceptualizing each step with a mental approximation or alternate computation to check errors, the entire process can be entered into a computer, even though its complexity may be such that no individual can conceive of all the steps at once unless he has very special training and talents.

As this situation came to be realized there was a rush of bright, mathematically-minded individuals into this new field who entirely outflanked and vastly impressed many of the older decision makers whose educations had been completed in the precomputer era. There was a quantum jump in the effectiveness of many kinds of analysis. The use of computer analysis to mathematically model pollution became increasingly influential. In particular, administrators and the courts were impressed by the specificity and apparent authenticity of the models; they gave clear-cut answers that allowed no ambiguity in interpretation. However, the glamorous new techniques brought some abuses with it. Some of the models have been based upon totally inadequate data, and the complex scientific relationships involved may have been inaccurately represented. Errors are all the more serious because they are so hard to detect behind the esoteric jargon employed. Among the many sources of error in modeling, perhaps the most frequent include misunderstanding of the nitrification cycle, the use of low-water steady state modeling to represent the worst pollution conditions without consideration of storm runoff and unrecorded pollution, and failure to consider the effects of future increased water supply withdrawals. Also, due to the interest in modeling, analytical emphasis has been too often confined to BOD and the oxygen cycle, which can be modeled readily, with too little attention paid to adverse effects of excess sediment, nutrients (which encourages eutrophication), heavy metals and other toxics. Modeling of eutrophication processes is a particularly difficult process to accomplish, and modeling approaches have tended to overlook effects of storm runoff pollution, because it is a temporary phenomenon that does not mathematically fit the steady state DO-BOD models usually employed.

We need more and better mathematical models, but we must remember that such models are grossly misleading unless they are based upon adequate data and designed with a clear understanding of the processes involved. Furthermore, the few scientists who have a broad knowledge of the biological effects of pollution are too busy to become engaged in the routine administrative rulings required for environmental impacts or waste load allocations.

Summary of Principles

The proper principles of contemporary water quality planning can be stated with some confidence. There still exists a national consensus that gross environmental pollution of our waters must be stopped, and the Water Resources Council is right in setting up environmental quality as a basic objective of planning, to be pursued concurrently with another basic objective, economic efficiency. (The national goals of the 1972 Act are logically unsupportable, in that they set up an objective of water pollution control virtually without reference to its economic costs.) Because with our present knowledge we cannot quantitatively evaluate the benefits of improved water quality, our approach to planning must be to set a desired environmental goal, and to outline the various alternatives by which that goal can be reached. The least costly of such alternatives is the preferred method to use to reach that goal.

If the costs for achieving the desired environmental goal appear to be politically unrealistic or out of line with the benefit expected, another less ambitious goal should be studied, and the least costly alternative to reach it should be pursued. The decision between the two alternative plans then remains to be made, in accordance with the multiple-objective approach, by balancing the extra costs of the original goal against its anticipated environmental advantages. Ultimately this is a political decision, since there is no scientific basis for such a choice, given existing knowledge, but the planning process must not obscure the fact that a decision is to be made.

The EPA procedures for areawide planning under Section 208 very clearly affirm this type of approach, in principle. However, it is not yet clear that these principles will in fact be applied. Some of the most emphatic and decisive language of the 1972 Act is that which sets up the national water quality goals and waste treatment standards as apparent absolutes, without reference to any planning processes. There is no language in the Act which specifically says that the results of the Section 208 plan will ever be allowed to modify the goals and standards specified in the Act. The present procedures of the EPA seem to indicate that facilities planning will not be modified by areawide planning, except perhaps in some future generation of facilities plans. Therefore, it seems that although there is really no question as to the proper principles, there is a very real question as to whether they will be applied seriously or merely used as window dressing, while the real decision making continues to be made otherwise.

7
Water Quality Control
in Other Countries

Politically and culturally, citizens of any large country tend to be somewhat myopic and the United States is no exception. Our country is so large, and so remote from major competitors and from the original sources of our culture, that we pay disproportionate attention to our own regional and national affairs and often ignore even very obvious lessons from abroad. Our national pollution control policy, adapting to our federated structure, has been left with obvious inefficiencies and incongruities. These organizational difficulties are particularly acute in major metropolitan areas. Therefore, it should be valuable to consider briefly how such problems are being handled in countries of Western Europe which are faced with very similar situations.

Of course, with our different political institutions, the organizational patterns of another country cannot simply be adopted in the United States, but the analogies should at least be suggestive. The idea of turning to western Europe for ideas in the water pollution control field was developed by U.S. analysts in the late 1960s.[1] At that time, certain German regional organizations for water supply and pollution control were already long-established, but their French and British counterparts were just developing. Since that time, they too have made major progress.

The Ruhr Region of Germany

The Ruhr region has long been famous as the center for heavy industry in Germany, starting with coal and steel production. It contains some forty percent of West German industrial capacity and a population of ten million people in an area of 4300 square miles, about one-third the size of the Delaware River Basin.[2] Starting in 1907, a series of water management districts, the *genossenshaften*, were created by special laws, with broader functions than exist for water organization in other parts of Germany.[3] Although the districts differ in detail, they essentially provide for centralized planning and administration of both water supply and water quality for each of the basins or sub-basins of the region. See map, Figure 7-1.

The Ruhr River is the largest of the five rivers serving the region, and its water quality achievements are extremely creditable. In spite of the large amount of industry in the upper basin, and the passage of the river through the City of Essen, which has a population of approximately one million, the entire

Figure 7-1. The Ruhr Region, Germany

river is kept suitable for fishing and general recreation, including contact sports. Within the city, the river is impounded into a shallow lake, and along the banks are shady and much frequented walks, cafés where bands play in the summer, and well used bathing beaches. The waters have many small sail boats, skiffs and the ubiquitous *faltboten* ("folding boats"). The Germans are practical people who have always developed their industrial potential, but they also love the outdoors. With true Teutonic efficiency they organized, even prior to World War I, to make sure that both of these interests would be well served.

Legally there are two organizations in the Ruhr Valley, the *Ruhrverband*, which controls water quality, and the *Ruhrtalsperrenverein*, which is responsible for water supply and the operation of the storage reservoirs. The two organizations, like others of the genossenschaften, have directorates constituted of representatives of municipal and local governments, coal mines, utilities, and industries. Since the entities served by water supply from the Ruhr are not identical with those whose effluents discharge into the Ruhr, and since the budgetary importance of the two functions varies between entities, the two organizations have different memberships, different budgets, and separate operations divisions. However, they are housed together, have the same managing director, and the same administrative staffs. (Much of the information on these organizations was given the author by Klaus P. Imhoff, now chief engineer of the Ruhrverband.)

The integration of water supply and water pollution control operations in the Ruhr Valley is most marked as concerns operation of the reservoirs. As noted above, United States water pollution goals are established as though water supply withdrawals and low flow augmentation did not exist. Not so in the Ruhr. The Ruhr water resource is known to be too important to waste by adherence to arbitrary rules. The reservoirs function primarily for water supply, but storage is accumulated mainly during high flow periods. During critical summer low flow periods, when dissolved oxygen descends to undesirably low levels, the reservoirs may release part of their stored water as low flow augmentation.

Further, at critical low flow periods the river is artificially aerated, disregarding a constraint which appears to be absolutely unassailable in the United States. When the dissolved oxygen in the upper part of the impounded river at Essen fails as low as 3.5 mg/l, a strange and awkward looking craft, decked with signs warning off swimmers, is towed out to a mooring place near the upper end of the pool, a diesel engine is started and two huge impellers throw foaming torrents of water out in all directions. This is a mechanical aerator. See Figure 7-2. A less conspicuous aerator is employed a few miles downstream. Air is bubbled up through perforated and weighted tubes which have been unrolled upon the bottom. A third type of aeration device is activated at the low hydroelectric dam which is located at the lower end of the pool. This device introduces air bubbles into the upper portion of the turbine draft tube, which operates under vacuum.

Aerators of these types are not unknown in the United States. In fact, the Ruhrverband staff were first encouraged to use them by reports of a field experiment on a mechanical aerator performed by the Chicago Sanitary District. Since then, a fairly extensive EPA demonstration project and several related OWRT research projects have shown the efficiency and practicability of such devices. Also, there are possibilities of mobile oxygenator systems which might be applied to large navigable rivers.[4,5,6,7] See Figure 7-3. Such studies have confirmed that, beyond a certain point, improvement of dissolved oxygen in rivers is much more economically obtained by instream aeration than by advanced waste treatment of point source wastes.

There are certain other features of the Ruhrverband approach which are of major policy interest to us. The first is the system of effluent charges. Industries in the Ruhr are given somewhat flexible limits of permissible waste load, and are charged a fee depending upon the amount of the loading actually discharged.

Source: Courtesy The Ruhrverband, Essen, Germany.

Figure 7-2. German Mechanical Aerator

The charge is determined by a rather complicated empirical formula, involving the oxygen demand and the toxicity of the wastes. The effect of the effluent charge is to provide a financial incentive to each industry to reduce its wastes by any means, including process changes. This system was of great inspiration to Allen Kneese, but his argument in favor of an effluent charge system has become somewhat less cogent by reason of new approaches to effluent loading limitations which, if properly administered, can accomplish much the same result.[8]

The second point of major interest is the treatment of toxic wastes. Certain industries, particularly electro-plating, generate wastes of high acid or cyanide content, or other toxics, which are difficult to remove by normal treatment processes. Two alternatives, other than normal treatment, are made available in the Ruhr. The first is disposal of waste waters in the Emscher River, which is lined with concrete and in effect turned into a huge trunk sewer. It not only conveys wastes from the highly industrialized Dortmund area in the headwaters, but also from industries in the adjacent water sheds, upon the payment of the

Figure 7-3. U.S. Mechanical Aerator in Passaic River

appropriate effluent charge. Near its mouth, before it discharges in the Rhine River, the entire flow of the Emscher River is treated; and the effluent charges reflect the cost of the channel maintenance and of this treatment.

The second alternative for industries in the Ruhr Valley with highly toxic wastes is to bring the wastes for treatment to the *Entgiftung* ("detoxication") plant, operated by the Ruhrverband. Transportation is usually in drums or cylinders. The wastes are analyzed, and specially treated so as to neutralize them as far as possible. Then the dewatered sludge is buried in an area whose geology can isolate it from any aquifers. Other less toxic waste water which is discharged into the Ruhr River is monitored to make sure that toxic materials have been isolated and disposed of in one of the two approved ways rather than being flushed into the river.

No one who studies this Ruhrverband system and visits its carefully planned, clean, and well operated facilities can fail to be struck with the general efficiency of the operation and the care taken to minimize environmental impacts. Waste treatment plants visited in the Ruhr Valley were generally nonodorous, and were carefully landscaped. The operating facilities were built at low elevations and were surrounded by banks covered with shrubs and small trees. In this way, adverse reaction to the locations of the plants is greatly reduced.

Great Britain

The regionalization of water supply services in Great Britain was initiated during World War II and continued over almost thirty years, whereby some 1300 separate water supply systems were reduced to less than 200 systems. While water supply systems in the United States serve an average of 5000 people each, the average served in England and Wales in 1974 was well over 250,000 people.[9] More than ninety-seven percent of the total population of England and Wales is now served by public water supply systems.

The idea of regional management of water pollution control in Great Britain became popular much later. In 1963, legislation was passed creating a system of twenty-nine river basin authorities covering the entire country. These authorities had responsibility for planning and regulating both water supply and pollution control, but had utility functions in only isolated cases.[10,11] In April 1974, ten enlarged and strengthened regional water authorities (RWA) took over virtually all responsibility for water from the twenty-nine previous water authorities, 187 water supply companies and about 1400 local authorities which had handled sewerage and sewage treatment.[12] The Regional Water Authorities have boundaries generally following river drainage basins, as shown in Figure 7-4.

The operational or utility functions authorized for the RWAs go far beyond those of the Ruhrverband. The financing policy will impose a single residential charge to cover water supply and sewage disposal services. Charges will also be made for discharge of industrial wastewaters into municipal sewerage systems, for discharge of industrial or agricultural purposes and for recreational uses of water. The aim is to make each regional authority self-supporting, with dependence upon central government only for loans for capital construction. Under such a system there is little likelihood of extravagance.

The RWAs, in addition to water supply and pollution control, have responsibilities for the use of inland waters for recreation, land drainage and flood prevention, and for fisheries in inland and coastal waters.[13]

The governing bodies of each RWA consist of a majority representing local authorities, with the remainder appointed by the government. Governmental responsibility for the RWAs rests mainly with the Secretary of State for

Source: R. Toms in "Urbanization and Water Quality Control, American Water Resources Association, 1975. Copyright by the American Water Resources Association. Used with permission.

Figure 7-4. Regional Water Authorities in Great Britain

Environment and for Wales, with some shared responsibilities for policy formulation and for land drainage and fisheries exercised by the Minister of Agriculture, Fisheries and Food. General planning for water resources is the responsibility of the Department of the Environment.

Immediate advantages of the reorganization, as seen by Ronald G. Toms, of the Wessex Regional Water Authority, include improved technical supervision over the smaller waste treatment plant operations, loyalty of inspectors to the system rather than to the municipality which hired them and which they were called on to inspect, better quality control of potable waters, including possible reuse of treated effluents, and easier planning for sludge disposal.[14] From an American viewpoint, it seems that relatively little attention is being given to consolidated ("regional") waste collection and treatment schemes, through regional plants and trunk sewers. Presumably such long range capital intensive changes will come slowly, in view of the economic condition of Great Britain. For the time being most of the water supply and sewage treatment operations are still being carried out by local authorities as agents for the RWAs.

A new "Control of Pollution Act" was passed in 1974, which was intended to strengthen the new RWAs, but it has not as yet been implemented. The act will combine legislation covering ground, water, air, and noise pollution, and will extend water pollution control to cover all discharges—including those directly to the sea—and also pollution of groundwater due to the use of the overlying ground. However, it also will remove restrictions on public availability of information, and will allow suits by private parties for noncompliance with the law. As a result of these idealistic provisions, the RWAs would be exposed to numerous lengthy and expensive law suits, because of the many waste treatment operations which they have just incorporated. Accordingly, the authorities are continuing to operate under the acts of 1951 and 1961, which places them in the somewhat anomalous position of being both the largest dischargers of effluent and the pollution control authorities in their respective areas.

It is, of course, too early to say just how well these new British systems will function, however, the experience under the previous river authorities has made it apparent that the combination of water supply and waste water treatment is a useful one. Moreover, a regional operation in which representatives of the local authorities can influence standards and budgets, is a safeguard against the fiscal unreality of our U.S. system, where national goals and deadlines were established without regard to the costs involved, and cannot now be met.

An example of the benefits of better scientific control of water pollution comes from an experience with damaged tomato plants in Essex, a county in southeast England. Investigation showed that the growers involved were all using water withdrawn from the rivers for irrigation, in which there was trace contamination of the chemical TBA. Tomato plants are sensitive to TBA at a concentration of only one part per billion. Investigation indicated that a recently initiated water diversion scheme transferred water from the Great Ouse watershed by pipeline. The source of the contamination was ultimately traced back to a plant producing pesticides on a small river. This plant had previously not been suspected, as it had a highly sophisticated treatment operation which included the use of activated carbon. The very small concentration of chemical which escaped to the effluent was not toxic to other plants grown in the Great Ouse watershed; accordingly the effluent had been considered satisfactory.

The Thames Estuary, which drains the city of London and much industry besides, has been notoriously polluted for several hundred years, however, the condition has been improved, as a tour by boat on the Thames River reveals. The estuary through the city, while turbid and lacking fish, is at least not odoriferous; above Hammersmith Bridge the water is better—there are popular rowing clubs, and eights, fours and sculling boats are abundant in summer; finally, when the boat passes the head of tide, the scene changes drastically. The nontidal Thames River, in spite of considerable urban populations and industry, is suitable for fishing, and most of the houses adjacent to the river are beautifully landscaped with gardens sloping down to the river. Towpaths along the banks (originally used to tow barges with horses) are now popular promenades. A short distance below Oxford (a sizeable city), a portion of the river is diverted for public swimming. The water is rather turbid and probably would not pass U.S. sanitary standards, but pollution control must be fairly effective to permit it at all. This section of the river is also used for boating and to train perhaps eighty university and college crews, in spite of the fact that the river does not average over seventy yards in width. The British method of pollution control is quite different from the German, but the great interest in attractive river environments is equally apparent.

France

In France, by 1964, it was obvious that a more concerted attack upon the water pollution problem had become necessary. Pollution was accentuated by urban concentration and industrialization, and 2500 kilometers of the larger rivers were considered to be severely polluted. Accordingly, in December 1974 a basic law was passed which may be summarized as follows:[15]

1. Water problems are allocated to one of six large river basin regions for solution. (See Figure 7-5.)
2. The cost of providing each user with the quantity and quality of water he needs is divided among the users, considering the benefit each derives and the disadvantages which his use creates for others.

Before considering the organization of the new agencies, it is necessary to understand the system of planning in France. Planning is conducted on a

ARTOIS PICARDIE

Douai

SEINE NORMANDIE Metz

Paris RHIN MEUSE

Orleans

LOIRE BRETAGNE

Lyon

RHONE

ADOUR GARONNE

MEDITERRANEE

Toulouse

CORSE

0 1 2 3

100 Km

Source: P.F. Tenière-Buchot and F. Fiessinger in "Urbanization and Water Quality Control, American Water Resources Association, 1975. Copyright by the American Water Resources Association. Used with permission.

Figure 7-5. The Six River Basin Regions in France

national basis, and results in the adoption of a "French National Five Year Plan of Economic and Social Development." This includes a general land use plan, and a transportation and utilities plan for each region. A portion of such a plan is shown as Figure 7-6. Although the French government does not go as far as Soviet Russia in directing new investment for housing and industrial and commercial development, the areas in which the entrepreneur may expand are spatially limited. Very extensive areas are designated as forests, parks, potential recreational centers, and *paysages de qualité*, which are presumably to be safeguarded against undue encroachment. While the plan is not an absolute control, it does have powerful influence: although local administrative agencies have the power to allow deviations, they may risk the displeasure of some powerful ministry by doing so. In addition to these five year plans, much longer range plans are prepared which serve as a general guide, and in turn influence the specifics of each five year plan. A similar approach to planning could not very well be incorporated into our present United States institutions because of the considerable power of the states. (However, it is noteworthy that a bill intended to give the federal government extensive functions in land use planning came fairly close to passage in 1975.)

It was not until 1967 that the new water organizations in France were actually created.[16] In each basin there is a committee composed of one-third government representatives, one-third local representatives, and one-third representatives of water users. These committees have considerable influence over the basin agencies, which are the administrative agencies for implementation of approved plans.

The committees (and the agencies) operate under various influences of seven different governmental ministries, as well as of the regional prefects. Ministers appoint their respective representatives on the committees. The local representatives are mayors, county councillors and presidents of trade associations, appointed by the county councils. The users represent industries, private water companies, water transportation companies, fishermen, farmers, tourist agencies and miscellaneous experts, all appointed by various nongovernmental organizations. Each committee elects a president and vice president from among its nongovernmental personnel. The agencies must consult the committees on all proposed programs. No program for improvement can be financed without committee approval of the program and also of the financial means to implement it.

The national government outlines the programs as a whole and subsidizes certain projects. The basin agencies define the objectives on the basin level, draw up specific proposals to correspond to the programs, and implement them financially. The current programs, which cover the five year period 1972-1976, include two basic objectives: the development of water resources and the fight against pollution.

Basin agencies provide financial aid to communities for construction of sewage treatment works, with subsidies of twenty to thirty percent of capital costs. Another thirty to forty percent may be provided by the central government, leaving the local community to finance the remaining 40 percent, which they may borrow at low interest rates. Under some circumstances, the agencies also may give subsidies and loans to private users who spend money on treatment plants or modify their processes in order to save water. The subsidy may vary from forty percent to eighty percent of the cost of the works. Financial aid is also given for good operation.[17]

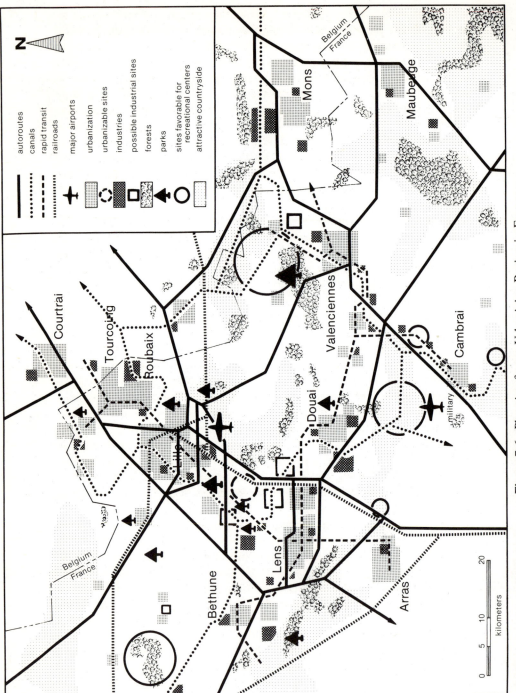

Figure 7-6. Planning for an Urbanizing Region in France

Legend:

- autoroutes
- canals
- rapid transit
- railroads
- major airports
- urbanization
- urbanizable sites
- industries
- possible industrial sites
- forests
- parks
- sites favorable for recreational centers
- attractive countryside

Mons

Maubeuge

Belgium / France

Courtrai

Tourcoing

Roubaix

Valenciennes

Cambrai

Douai

military

Lens

Bethune

Arras

Belgium / France

0 5 10 20
kilometers

About twenty-seven percent of the revenues of the agencies come from water withdrawals and seventy-three percent from effluent charges (pollution fees), with considerable variation in this rate from basin to basin.[18] The effluent charges are based upon an empirical formula (somewhat resembling the Ruhrverband formula) which estimates pollution loading by taking into account the suspended solids, oxidizable material, salinity, and toxicity. The charges are higher in those regions where the programs are the more costly.

The total size of the national program for the five-year period is not very large by U.S. standards. The total estimated returns from water withdrawal fees and effluent charges will only amount to $490 million, or slightly less than $100 million annually.[19] The estimated total capital costs of subsized new treatment plans for the five year period is estimated to be about $528 million, of which the central government subsidy would be $180 million. The expenditure for sewerage systems for the five-year period would be about 92 to 125 million dollars, of which the central government subsidy would be ten percent.[20] The above estimates indicate a total expenditure in connection with the government subsidized program of less than $150 million annually in 1970 dollars, for a population of one-quarter that of the United States. The inference may be drawn that the subsidized pollution control program is proportionately much less in scope than that of the United States.

Although the basin committees are largely composed of nongovernmental personnel, this is not true of the agencies. Each agency has an administrative council of twenty members, of whom ten are from government, five are from local agencies and five are water users, chosen from among basin committee members. The head of the agency is a director, appointed by the government.

Coordination on a national level is provided by an Interministerial Water Commission, headed by the Secretary of State for the Environment. The commission approves the various agency programs, with the advice of respective basin committees. There is also a department of water problems, a part of the Ministry of the Quality of Life, which serves the Secretary of State for the Environment in his supervision of the basin agencies. Figure 7-7 illustrates the established channels in summary form. To U.S. eyes this organization seems unnecessarily complex, although admittedly our own organization for pollution control is hardly a simple one.

There is a considerable role for private water companies in France. Private distributors account for fifty-five percent of total water sales, and they are continually expanding. The largest, the *Société Lyonaise des Eaux* and its subsidiaries supply approximately eight million people with water, and provide sewage disposal for about two million. It is apparent that the proliferation of very small water supply companies, which progressed so far in Great Britain before consolidation began and is still very prevalent in the United States, did not occur in France to the same extent.

The character of the developing water pollution control programs in France

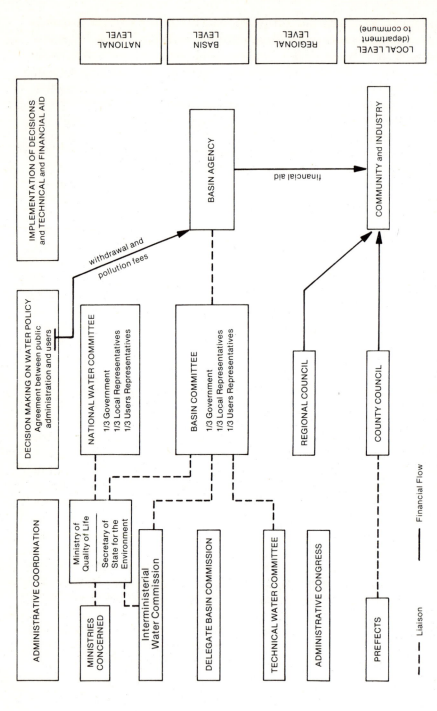

Figure 7-7. Public Water Organization in France

is still not as clear as that in Germany and Great Britain. The channels of control are much more complex, with a much stronger dependence upon the central government. In only one respect is the French picture superior: effective overall planning will provide the water pollution control planner with the necessary direction for his projections and programs. This direction cannot be provided with anything like the same assurance in Great Britain or in the United States.

It is also certainly true that in France plans should be able to take into full account the water supply as well as the water quality pictu e. In this respect all three of the European countries have established the necessary integration. It is also apparent that French water quality standards will be set with substantial influence by regional authorities and users who are responsible for paying for them. Thus if one region of France is more concerned about environmental quality than is another region, it will set higher standards and pay the higher charges to obtain those standards. This is also the case in Great Britain and Germany. Is this really a bad thing? Just why has it been accepted in the United States that standards of treatment everywhere should be exactly the same? France, England and Germany are countries which have faced hard economic times and diminishing natural resources, but by and large these countries remain environmentally more attractive than ours. They face the same problems of crowding in metropolitan areas, and of overuse of their rivers. We should carefully consider whether there are not some lessons for us in their experiences.

Water Quality Standards in Japan

It is interesting to compare our water quality standards in the United States with those in Japan. According to a publication of the City of Osaka, Class A waters, which are fit for trout, for bathing, and for water supply (without special treatment) should have a BOD not over 2 ppm, and dissolved oxygen of 7.5 or over.[21] Class B waters, which are suitable for water supply only with special treatment and for fish such as salmon, should have a BOD not less than 3 and DO of at least 5. These standards are higher in some respects than corresponding standards of the United States. However, the schedules for implementation are more realistic than ours. As compared to our goal of fishable and swimmable waters throughout the entire United States by 1983, the immediate goal for most of the rivers within the city of Osaka is Class E, which calls for a BOD of 10, DO of 2, and "no rubbish floating," with no limit on suspended solids. No definite time limit is set for the achievement of these immediate goals. The metropolitan area of Osaka had a population of 6,092,000 in 1975, a great deal of industry, and a complex network of rivers and canals within the city, many of which have been extremely polluted. Therefore, it will require a major effort to achieve a minimum standard even of Class E.

8

Water Pollution Control Strategy and Combined Planning

Environmental Objective

We have seen in Chapter 2 that the need of mankind to have access to natural environments is basic to his nature, this need being apparently largely genetic, but also culturally developed and influenced in the early years of infancy. The attempt to find religious or ethical sanctions for the preservation of wild species in and for themselves has not been successful; plants and animals and their natural surroundings have importance to man only insofar as they are of interest. For most persons, this interest is considerable. Although, when faced with conditions of economic hardship or danger, men have always been willing to accept a considerable amount of environmental degradation, a national consensus arose in the United States within the last decade to the effect that the environment was generally too degraded, that water, air and solid waste pollution were at unacceptable levels in many of the developed portions of the country, and that corrective action to protect the environment was long overdue, and should be expedited. The emotional fervor of the environmental movement has now declined, but there is still a great deal of solid support for such programs.

There are three main thrusts to the environmental movement. They are (a) preservation of natural species and natural lands and waters, (b) improvement of man's living and recreational facilities, and (c) enhancement of public health.

With respect to any proposed public action, the economic advantages and disadvantages of each alternative must be considered, as well as the environmental effects. The latter can be considered in terms of providing generally improved conditions for life, work and recreation, and also in terms of any special advantages or disadvantages, of a cultural, aesthetic or ecologic nature, which may be of interest to a more limited public. Consideration of economic advantages or disadvantages may utilize numerical indices and other weightings, but only if used with recognition of their subjective nature. Some sophisticated systems analysis approaches now being elaborated give a fallacious impression of having solved the entire problem of evaluating environmental quality in money terms. Although the complete solution is still beyond our evaluation, some useful partial evaluations are available and many of the relevant environmental parameters can be described and some can be evaluated in money terms. As regards others, physical standards can be set for achieving given social objectives, such as providing bathing facilities or nature trails for a given population. In principle, it is quite clear that the main objective is that of providing (or maintaining) an outdoor environment suitable for the enjoyment and psychic health of everyone. The complete environmental program should include a balance between man and nature. The environmental objective is not a categorical imperative, but is subject to tradeoffs with other social goals and with potential economic advantages.

Dr. Ruth Patrick has stressed the necessity for a balanced approach to the maintenance of ecosystems and for flexibility in the application of criteria.[1] She feels that set standards should take into account the natural characteristics of surface waters, the type of ecosystems present, and the functions that various organisms carry out (e.g., spawning) in the system which we are seeking to protect. The degree of pollution control applied to heavily used stream reaches should vary according to the assimilative capacity of the stream, as long as the pollutants are easily assimilated or dissipated. However, she would like to see natural wetlands preserved, and some wilderness areas held inviolate by man with others being managed for correct use without abuse.

In most areas it is inherently impossible to preserve natural species in their original habitats and simultaneously make the same area available to man for occupancy and recreation. Such conflict in aims can only be reconciled by remembering that it is the interest of mankind that constitutes the ultimate sanction, and that viewpoints vary considerably. The planned environment in metropolitan regions must include some completely wild areas, preserved for study and as a remembrance of the past, but also must include recreation grounds with plants, animals and water, which may be related quite differently from natural ecosystems. The children's zoos and park lakes with ducks and goldfish can play an important social role in maintaining the psychic health of thousands of urban dwellers. The extensive facilities of Jones Beach on Long Island, which offer healthful outdoor experiences to millions, were built on lovely, unspoiled barrier beaches, where bird watchers and other lovers of solitude, and an occasional Coast Guard patrol, once found a virtually primeval ecosystem. Nostalgic regrets for that natural environment (which the author experienced as a boy) do not correspond to a realistic view of our society. The ethic of the metropolitan environment must give first place to man, and the unaltered wildlife sanctuaries have to be located so as to be consistent with more immediately pressing social objectives.

The environmental objective as discussed above was not clearly expressed in the great environmental movement (in particular, the incompatibility of preservation and public recreation goals in any one locality was carefully avoided). The movement quickly turned into a double-headed drive, first, to halt construction of as many large public works as possible, and, second, to eliminate pollution of waterways particularly by industries. Thus the emphasis was focused into an attack on clearly defined targets, and was to some extent antitechnological. The halting of public works was accomplished by means of environmental impact statements, which offer endless possibilities of delay, with the proponents of any project being obliged to spend large sums of money for analyses and reports, but with no very clear channels for resolving the problems in the end. The halting of industrial pollution (and to a lesser extent municipal pollution) was to be accomplished through the Pollution Control Act of 1972. Large industries were the main immediate target. This was politically a viable and very effective posture.

The Basic Problem

As a nation we now find that the combination of the 1972 Act and the National Environmental Policy Act are not serving our purposes fully. We really must find a better way of handling our water pollution problems. The surrogate objectives of halting public works and of insisting on high degrees of wastewater treatment have real advantages, but they are being carried on too inflexibly, too indiscriminately, and they are not making enough progress towards improving the environmental conditions. Our waters are almost as polluted now as they were when the act was passed, and the prospects of major improvements within the few years specified in the Act are now seen to be illusory. Moreover, the monetary expenses (which must be understood to include manpower, energy and raw materials) are astronomical.

As we saw in Chapter 1, the full implementation of the 1983 goals would cost hundreds of billions of dollars, sums far in excess of what can be provided. It is apparent that only a fraction of the program could possibly be implemented by 1983. Therefore, priorities are required to determine which of the various parts of the system are most important.

There is serious doubt that the pursuit of the goals of the 1972 Act, even if they could be achieved, would lead to commensurate environmental improvements. Progress at current levels of funding would be more apt to result in a patchwork of treatment plants, some built to extremely high standards, and many others, perhaps the majority, hopelessly inadequate. In the opinion of most professionals, this program would entail a great waste of materials, manpower and energy. It is not an appropriate goal upon which to base planning or individual facilities, since unless flexibility is built in, it is apt to result in expensive overdesign.

The provisions of the law requiring uniform national standards of "best practicable" or "best available" means of waste treatment specified for 1977 and 1983 are not enforced literally, because to do so would be as unfounded as to prescribe that all highways in the United States be built to the same standards. The mind revolts at such a suggestion for highways. A highway traffic survey would recommend that our main interstate highways be built with six or eight lanes of heavy duty concrete pavement, and county roads with two lanes of light bituminous material. Similarly, there is no logical reason for applying the same standards of pollution control coast to coast, regardless of the local situation of the effect produced. Uniform standards would serve the interests of the crowded metropolitan areas by discouraging industry from choosing other regions for development. However, we cannot afford this type of parochial thinking. It would be far better for our society as a whole if new industrial plants were located where they would do the least environmental damage. Moreover, the idea of applying "best available" or, worse yet, "no discharge" standards of treatment to the return flows from irrigated agriculture in the West is completely unrealistic. Some very different approach will be needed.

When first enacted, criticism of the national water pollution control program was restrained, but now criticism comes from all directions, in a steady roar. Although there is great diversity in the views expressed (and some of them are obviously selfish, anti-environmental, or simply confused), there is a substantial amount of dissatisfaction with major aspects of the program. However, very few critics have tried to suggest a better general approach to water pollution control. This is what has to be attempted in this chapter.

Water Resources Council Procedures

It has been suggested that the official guidelines supposedly applicable to all federal water resource planning, the famous "Principles" of the Water Resources Council should be applied directly to all water quality planning.[2] This would call for planning to be directed towards twin national objectives of environmental quality and economic efficiency. It is stated in the "Principles" that where economic benefits cannot be established for environmental quality, some other quantitative means should be devised. The attempt to do this has lead to a variety of index approaches, units of recreational carrying capacity, etc., which are at best subjective. They represent a blind alley with respect to other environmental considerations, such as aesthetic value and preservation of ecosystems. With our present knowledge we cannot quantitively evaluate the benefits of environmental quality in the same way as the benefits of other programs, such as flood control or water supply. The technology to apply the same planning approaches to water quality is simply not available at this time.

Some of the advocates of applying the "Principles" to water quality planning are theorists, who realize the present lack of planning technology, but who hope that a "crash" federally supported research program would quickly produce the heeded technology. Others who say they agree with this approach are thinking of something quite different. Under the "Principles," water resource planning may incorporate restraints or requirements imposed by law, so that treatment required by the nationwide standards of PL 92-500 could simply be incorporated into the plan as a starting point and other elements then developed around them. If so, the application of the "Principles" would be purely nominal. This approach would affect a legal reconciliation of the two programs, but it would fail to achieve the advantages of a true combined approach, as discussed above. It would be misleading to say that the procedures of the "Principles" were being applied, if that was all that was meant.

Two Panel Views

It is interesting to note that when the present PL 92-500 was under consideration in December 1971, a distinguished panel of experts was convened by the Director, Office of Science and Technology, Executive Office of the President, to advise on major issues. Among the panel's first conclusions were the following: 1) the goal of eliminating discharge of all pollutants "may be unrealistic as well as being undesirable from both the economic and environmental points of view" and 2) the use of national uniform water quality standards is an example "of arbitrary decisions that can impede an optimal strategy of water quality management." The foresight of this ad hoc panel was highly creditable, the membership distinguished, and the language clear and forceful. However, at that time the political winds of America were blowing so strongly in favor of unrestricted pollution control that such cautions went unheeded.

Another ad hoc panel of professionals assembled (without official sanction) at a national symposium of the American Water Resources Association (AWRA) in July 1975. Briefly, the conclusions[3] of this panel were that PL 92-500, while admirable in its aims, is unrealistic in its goals, arbitrary in its approach, and of doubtful environmental effectiveness. The 1975 panel cannot claim as much foresight as the earlier panel, since there is now abundant evidence of the faults of the present strategy, however, the political climate has now changed, and there is hope that the issues can now be discussed more rationally, and that professional views and accumulated evidence may now have meaning for the decision makers. A listing of the panel members and a statement of their recommendations is given at the close of this chapter.

Nonpoint Sources, Unrecorded Wastes and Urban Runoff

One of the reasons that the approach of PL 92-500 must be reconsidered is that while the treatment-only strategy applies only to the wastewaters from identified municipal and industrial sources, it is now generally acknowledged that in developed areas at least half of the pollution may come from unrecorded sources, either nonpoint sources or urban runoff. This includes not only BOD but suspended solids, heavy metals, nutrients and hydrocarbons.[4,5] Details are given in Chapter 4. Studies of this unrecorded pollution and urban runoff by various analysts, including those of the EPA, show that it would be wildly uneconomic to require treatment of all such runoff as it occurs. A lot of pollution comes roaring out with storm flows, building up to huge rates of loading perhaps a half-hour after the start of a heavy rainfall, and tailing off rapidly thereafter. Other pollution comes seeping out from leaky sewers, or is poured out surreptitiously by illegal dischargers. Combined sewer overflows are also of great importance. Moreover, all of these combined storm and unrecorded municipal loadings have a variable impact upon the main streams, not easy to calculate by low-flow steady state models. No one now maintains that a treatment-only formula, applied at Washington level, is going to solve these urban runoff problems. Various combinations of detention storage, segregation, treatment of certain fractions, and nonstructural means are apt to be required. Therefore it is absolutely essential that alternative approaches be systematically considered in areawide planning.

Basic Concept of Water Quality Control

Summarizing the discussion up to now, the present overall pollution control program of PL 92-500 is essentially a "treatment-only" approach, which sets standards of treatment at least nominally on a uniform national basis, and requires this treatment to be undertaken practically as an end in itself, regardless of the planning of our nation's rivers for other uses, regardless of economics, and in some respects, regardless of the environmental effectiveness of the program. It is true that in practice the EPA's actions are more realistic than the language of the Act would indicate, but it is totally unsatisfactory that EPA officials should have to distort the plain meaning of the Act's wording in the interest of common sense and justice. If the language of the Act is unreasonable it should be changed. We need to have laws that mean what they say, and officials who can be held responsible for enforcing them. A consensus appears to be building up among objective analysts that our national approach to water quality control should abandon the "treatment-at-all costs" approach and aim directly at improving the environmental quality of our rivers. The problem of course is to outline an approach to doing this. It is also necessary to consider how water quality planning can be included as part of a combined approach which will consider other program objectives, such as water supply, flood control, and public recreation.

Adapting the Approach of Section 208

If environmental quality is to be our basic goal (rather than waste treatment for its own sake) and if we are to consider the costs of achieving it (rather than assuming that any degree of waste treatment is automatically justified) we need a concrete basic planning procedure. Nothing like the benefit-cost approach or the index approach is practicable for this purpose.

One important basic action which has already been taken is the establishment of water quality standards nationwide. Some of these standards are unrealistically high and probably can never be achieved; but at least they provide a useful starting point. As shown previously, basinwide or areawide water quality planning is supposed to determine the most cost effective means of achieving water quality goals. A complete investigation and comparison of alternatives is to be carried out, and a determination made of the most cost-effective way of achieving the goal. This is undoubtedly on the right track. If some of the constraints imposed by EPA under PL 92-500 were removed, the procedures prescribed for Section 208 planning would be much more effective. Three major changes are suggested.

Change I: Alternate Goals

The first change suggested in the present Section 208 approach is to add an additional step. Whenever there is doubt as to the practicality or feasibility of achieving a given water quality goal, the costs of achieving a lesser alternate goal would also be investigated. For example, the basic goal for the Potomac River might be fishable, swimmable waters throughout its length. However, when full account is taken of the urban runoff from the metropolitan area of Washington, D.C., it may be altogether too costly or entirely impracticable to maintain swimmable waters adjacent to it. An alternate goal might maintain that the estuary be generally fishable, but at times allow the water in urban areas to be unsuitable for contact sports. This does not mean that the lower alternate goal would definitely be adopted, but only that the costs of both goals would be investigated and considered before a decision was made. If the planners worked out least-cost plans for several different environmental objectives, they could be entered into a decision matrix (Table 8-1).

Use of such a matrix approach, considering alternative solutions, would not avoid the problem of evaluating environmental quality; instead, it would drag the key decision out into the open for public discussion and consideration. That question is: to what extent can we afford to approve environmental quality at any given time? Under present procedures, that question is carefully concealed and avoided by the time tables and standards of the 1972 Act. From a matrix such as the one given above, the local, state and federal decision makers and the public would all know that it would cost $2,000,000 more annually to improve the estuary waters for general recreational purposes, and $2,000,000 annually above that to make them really clear and fit for swimming. The cost of treatment would be balanced against the costs of preventing pollution by surveillance and enforcement in urban and industrial areas.

Table 8-1
Decision Matrix (estuary)[a]

Goal	Least Cost Plan	Cost over next ten years, both government and private
A. Continuation of somewhat polluted water, at times odoriferous, but not septic at surface, only a few coarse fish. Shellfish inedible.	Secondary treatment of all effluents	$2,000,000 annually
B. Water suitable for boating fishing; but murky at times and not suitable for contact sports in urban areas. Shellfish satisfactory.	Secondary treatment of all effluents, land use restrictions on expansion of industry, storage and primary treatment of first flush urban runoff.	$4,000,000 annually
C. Water throughout the estuary suitable for all recreation including swimming. Water clear and clean.	Secondary treatment of all effluents, tertiary treatment of specified facilities, low flow augmentation, land use restrictions on expansion of industry, all construction, and commercial centers. Retention and secondary treatment of first flush urban runoff.	$6,000,000 annually

[a]This is not the Potomac River estuary, the costs of which would be far higher.

Change II: River Flows

Another major change from present approaches would relate to the flows of water in the rivers. At the present time, it is stipulated that projected drought flows of the river be considered as the basic conditions for water quality planning. However, this approach should be modified by considering possible changes in flow regime. Natural stream flows in developed river basins are usually greatly modified by consumptive use, interbasin diversions, and river regulation through storage reservoirs, and plans for the future should consider (a) probable consequences of further reductions in flow due to projected increased usage and diversion and (b) possible increases through low flow augmentation. Possibilities of improvement of flows should no longer be dismissed airily with the simplistic slogan "dilution is not the solution to pollution."

Change III: Use of Planning Results

The third major change in the areawide and basin planning is that the results of planning should really be used. We should have no more EPA injunctions that areawide planning should not interfere with ongoing or completed facilities planning. If the areawide planning shows that a given facilities plan will result in spending millions of dollars without commensurate return, that facilities plan should be changed. Mere administrative convenience and consistency with previous announcements is not sufficient reason for wasting large amounts of manpower, materials and energy. Moreover, alternatives other than treatment, such as river aeration, would be recommended for use where shown to be economic and not automatically given a pocket veto.

What has been outlined so far is an approach that could be used without major change in our governmental organizations or responsibilities. It is shown in Figure 8-1. Some enthusiasts for waste treatment at any cost will object that it opens up the key issues involved for too much public discussion. However, we really should have public discussion of the real issues involved. At the present time, prolonged and detailed public participation is involved in preparing each of the Section 208 plans. Such public participation procedures are little more than window dressing, if the key issues are decided upon through automatic formulas and interpretations, if these decisions are not to be changed even if shown to have been erroneous, or if the planning process is so managed that the key issues are not considered at all. The resources available for water pollution control in any given area are not limitless, and the issues involved are important enough to warrant a rational planning approach, and open discussion, followed by Congressional action in important cases.

Revision of Water Quality Standards

Adoption of a revised approach would no doubt result in statewide recommendations for many changes in our present water quality standards. Most of the changes would be for lowering the standards which would be an improvement from the present situation, where unrealistically high standards remain on the books for public relations purposes (and possibly for application against some helpless defendant), but without any pretense of general enforcement. However, in an effective system, there must be much improved means of making water quality standards mean something. What is suggested is that after the costs of alternative programs are developed through water quality planning, the state be allowed to submit a revised standard for the basin concerned for approval by the federal government. The EPA should send its recommendations for such changes to Congress, which should incorporate them into an authorization bill. (This is comparable to the procedure followed for authorization of projects of the Corps of Engineers.) Water quality control decisions for large metropolitan basins may involve a good many million dollars, and are of ample importance to warrant congressional action.

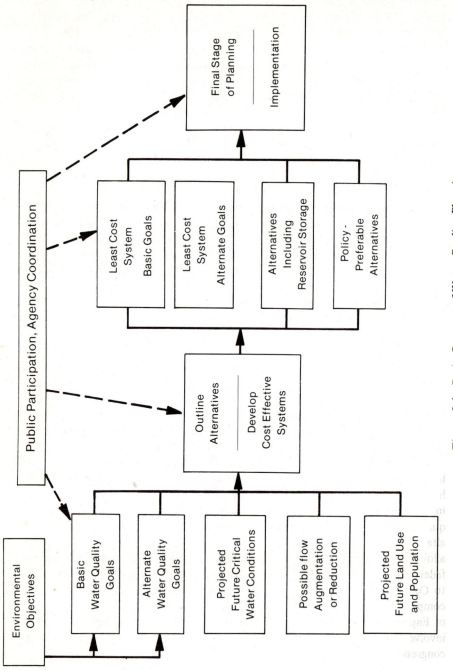

Figure 8-1. Basic Concept of Water Quality Planning

Incentives and Enforcement

Once established by congressional action, water quality standards should be enforced. It would be unrealistic to expect state enforcement as they only adopted the present standards under great federal pressure and the environmentalist influences of the times, and they may not take even revised standards much more seriously. Enforcement could be aided by an incentive system, but first we must consider the present system of incentives.

The EPA currently pays for seventy-five percent of the capital cost of municipal wastewater treatment systems, but nothing towards the operations and maintenance costs. This is inherently not a very good arrangement, as it leads to designing capital-heavy plants, in order to minimize the operations and maintenance costs which the community must pay itself. Moreover, once the subsidy for that particular community is given, it can no longer act as motivation towards continuing cooperation. If the federal subsidy, at whatever level considered appropriate, were divided between total capital and operating costs, to provide say fifty percent of each, the operating cost portion could then be used as an incentive to reach and maintain approved water quality standards. The subsidy should apply not only to conventional treatment plants but also to other facilities or operations such as treatment of urban runoff if that were economical. In any river basin, the state (or interstate) water quality control authority would outline the measures appropriate to achieve designated standards. The states have considerable difficulty in enforcing needed action in counties and small towns, and it's almost impossible for states to influence the large cities. Under a system as proposed, the fifty percent subsidy of operating and maintenance costs, which would be a large amount for major waste treatment systems, could be made conditional upon meeting established water quality standards. The EPA could withhold the subsidy, or part of it, for river basins whose waters failed to meet the standards, and the state could apportion the deficiency to uncooperative communities, based not only upon effectiveness of waste treatment plants but also overall effectiveness in reducing pollution from unrecorded sources and urban runoff.

Other incentive systems would be needed for industry. As previously discussed, either a waste load allocation system or an effluent charge system can be used to arrive at an optimum treatment program, representing the lowest total cost of achieving a given water quality goal. However, there are certain equity considerations which could only be met by effluent charges. For example, it might prove to be cost effective to require a very large industry to treat its wastes to a ninety-five percent level, a second smaller industry to treat to eighty-five percent, and some very small or specialized industries to give primary treatment only, or none at all, since their higher marginal costs would make treatment uneconomical. However, this approach would be inequitable, since it would not require contributions by various industries proportional to the respective pollution loadings discharged to the stream. Some system of effluent charges might be applied to industries and also to cities which allow pollutants from point sources to be discharged into the streams. The returns from such effluent charges could be used to defray the costs of monitoring the control of pollution, and also to pay for supplemental water quality control systems, such as river aeration, which are not chargeable to any one industry or municipality, but the use of which reduces the expenses for effluent treatment which otherwise would be required.

Environmental Impact Determinations

The results of water quality planning should not only be used to guide programs of wastewater treatment but also other matters related to water quality. In particular, such planning should provide the basis for estimation of environmental impacts. It is very inefficient to require many different companies and agencies on a given river to separately estimate the effects upon water quality and aquatic ecosystems of the various proposed developments for which they are respectively responsible. As brought out previously, power companies, industries, bridge and highway authorities are not experts on environmental conditions; they must rely upon consultants, whose findings frequently lack an adequate factual basis, and may be biased towards their clients. The planning of water quality should be conducted in such a way as to provide a basis upon which future changes due to construction or other activities could readily be estimated. In important cases, the federal or state agency itself should conduct or fund a study to determine what environmental effects should be expected by various important alternatives. If the agency is to exercise the power of decision, it would be better for it to conduct an objective study itself, and take the responsibility for it, rather than cautiously assuming an umpire's role between contending parties. The present system is very beneficial to consulting engineers, and it effectively insulates federal officials from being held responsible for inadequate studies; but it is not the best way to build up fully competent and reliable professional staffs in state and federal agencies.

Regional Sewerage Systems

Much more careful attention should be given to proposals for centralized waste treatment plants, designed to collect wastes from widely dispersed locations with large trunk sewers. The very real economies of scale made possible by large plants add to the advantages of greater ease of supervision by government agencies and the technical advantages of large plants. There was an initial rush of enthusiasm for "regional" systems, and some states seemed to believe that the adoption of such systems statewide was obviously both desirable and economical.

It only gradually came to be realized that the economies of scale in treatment may be counterbalanced by extra costs of long distance sewage lines. Some of the economies of scale were illusory: they were based upon comparing costs of one large new plant to several small new plants, but insufficient attention had been given the fact that some of the existing small plants were already operational and reasonably adequate.

When communities were offered the opportunity to combine in the new and supposedly preferable groups under combined state and EPA pressure, they often showed themselves surprisingly reluctant. For a variety of motives, communities seem to want to operate their own waste treatment facilities. There was usually no existing governmental organization ready to assume the management task in which the other communities had full confidence. It is not merely vanity that causes a community to hesitate to turn an essential service over to some extraneous group; it may manage the facility without due prudence and economy and forward the bills to the community to pay.

Also there were environmental problems. Some of the central treatment plants would have discharged huge volumes of effluent a short distance above water supply intakes, and the transmission of wastewaters downstream to central plants would leave certain tributaries almost dry during summer periods.

However, some less obvious indirect effects may be even more important. A study by the Metropolitan Washington Council of Governments showed that provision of sewer service has a more important relationship to the development of land than water service.[6] The location of sewer facilities creates a capability for development, independent of and sometimes in conflict with general land use plans or zoning. There are three important aspects:

1. location of sewer lines, pump stations and treatment plants
2. capacity of sewer lines
3. timing or staging of sewer lines

Spheres of influence of the lines extend to all areas within which connection is physically possible. Proposed open space within the spheres of influence needs special protection. Most economic development between the years 1960 and 1968 did occur within these spheres, in the areas that were studied.

It was once thought that land use zoning and limitation of the capacity of treatment plants could be relied upon to control development. However, the courts have increasingly declared invalid most zoning laws which were alleged to be discriminatory, and utilities may be forced to extend service to developers whether or not they wish to do so. Despite moves to the contrary, it seems that the most probable outcome of extending sewage collectors across county is that development will follow. Of course, in many communities such development is welcome, but in others it is not, and the adoption of a regional system is seen as contradictory to the planning objectives of a community.

These issues were exemplified by the communities around Princeton, N.J. on Figure 8-2. The main plant would be implemented under either a centralized or decentralized plan, with major connectors. However, the townships of Hopewell and Lawrence objected strenuously to the plan because of the unwanted development which the connectors were expected to encourage, and also the diversion of flows from the two brooks affected. The alternative is to provide two minor plants, designed for very high quality effluents, and also to improve two small package plants which are already in existence. Besides the engineering and economic comparisons between the costs and operating effectiveness of the two sets of plants, there is a key intangible. If after the plants are all built, future administrations of Hopewell Township prove unable or unwilling to maintain the zoning provisions upon which the plan is based, the sewerage authority as a whole would have to stand the extra cost of providing additional treatment capacity at the relatively expensive upstream plants, particularly the package plant, which would be very expensive to enlarge. With a single plant system, sufficient capacity in the trunk or interceptor sewers could be provided to handle all growth likely to occur. This problem is essentially a choice between land use philosophies. It must be solved by a plan agreeable to the various communities concerned before the matter can be settled.

It is apparent that only very detailed and comprehensive planning for each specific case can solve problems such as this.

Institutional Problems

Another difficulty with water pollution control planning is that the institutional framework for PL 92-500 is entirely different from that for other water resources planning. The Corps of Engineers carries out its planning through its district and division offices, which plan for improvements in river basins, including flood control and flow regulation, and recreational facilties for reservoirs. In contrast, the EPA provides funding for areawide planning by politically designated areas, which are often truncated portions of river basins. Moreover, the EPA does not generally carry out this planning itself. Therefore, any coordination must take into account two vastly different organizational approaches. In many ways, it would be preferable to create integrated river basin authorities, such as those in France, Great Britain and the Ruhr District of Germany, described in Chapter 7, which could conduct planning of water supply and water pollution control together. However, the systems of the EPA and the Corps of Engineers are probably too firmly established for such a change to be adopted generally in the United States. Basinwide approaches on interstate streams can only be accomplished by means of interstate compact, or by the

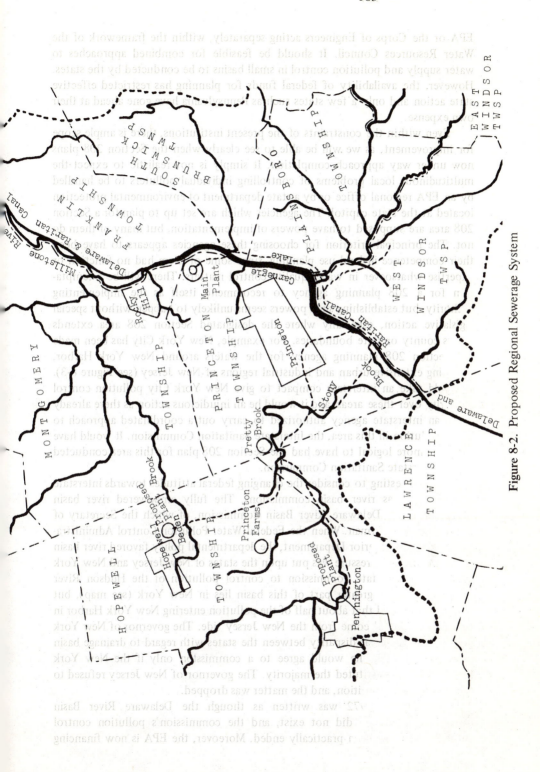

Figure 8-2. Proposed Regional Sewerage System

EPA or the Corps of Engineers acting separately, within the framework of the Water Resources Council. It should be feasible for combined approaches to water supply and pollution control in small basins to be conducted by the states. However, the availability of federal funds for planning has restricted effective state action and only a few states such as Pennsylvania have gone ahead at their own expense.

Even within the constraints of the present institutions, there is ample scope for improvement, as we will be able to see clearly when the Section 208 plans now under way approach completion. It simply is not realistic to expect the multitudinous local problems of controlling individual polluters to be handled by an EPA regional office or by a state department of environmental protection located at the state capitol. The agencies which are set up to plan for a Section 208 area are supposed to have powers of implementation, but many of them do not. The principal criterion for choosing these agencies appears to have been their competence in land use planning, since some of them had no function or expertise whatsoever in water quality control matters. There will be a temptation for a 208 planning agency to recommend itself as the implementing authority, but establishing such powers seems unlikely to happen without special legislative action, especially where the designated Section 208 area extends across county or state boundaries. For example, New York City has been made the Section 208 planning agency for the waters around New York Harbor, including extensive urban and industrial regions of New Jersey (see Figure 8-3). It would take an interstate compact to give New York City pollution control authority over these areas, and it would be an injudicious action as there already exists an interstate agency authorized to carry out a coordinated approach to water pollution in this area, the Interstate Sanitation Commission. It would have appeared more logical to have had the Section 208 plan for this area conducted by the Interstate Sanitation Commission.

It is interesting to consider the changing federal attitudes towards interstate agencies such as river basin commissions. The fully empowered river basin prototype is the Delaware River Basin Commission, of which the Secretary of the Interior is chairman. When the Federal Water Pollution Control Administration was in the Interior Department, the departmental policy favored river basin commissions, and pressure was put upon the states of New Jersey and New York to form an interstate commission to control pollution of the Hudson River Basin. By far the greater part of this basin lies in New York (see map), but forecasts indicated that about half of the pollution entering New York Harbor in future years would come from the New Jersey side. The governor of New York pointed out the great disparity between the states with regard to drainage basin area, and said that he would agree to a commission only if the New York members of it constituted the majority. The governor of New Jersey refused to accept a minority position, and the matter was dropped.

The Act of 1972 was written as though the Delaware River Basin Commission (DRBC) did not exist, and the commission's pollution control authority has now been practically ended. Moreover, the EPA is now financing

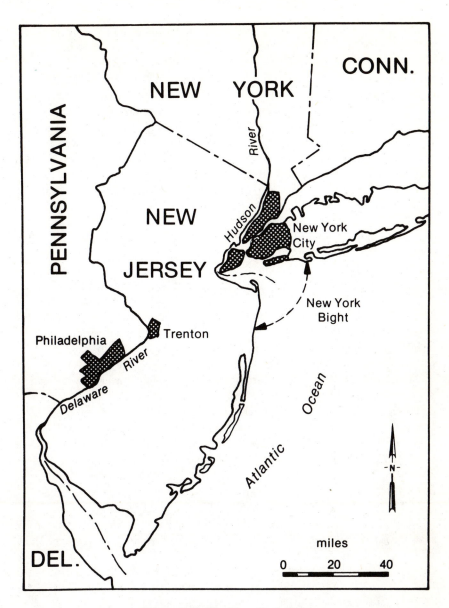

Figure 8-3. The New York Bight

areawide pollution control planning of key metropolitan areas of the Delaware Basin by another agency under Section 208.

It hardly seems in the best interest of the United States to maintain an interstate agency for comprehensive water planning and pollution control, and then rely instead on ad hoc and piecemeal areawide planning provisions. If the

DRBC needs to be modified to meet changing needs or reorganized so that it must plan in consort with other agencies, this should be done, but only if authorized with responsibilities commensurate with its statutory mission. We need regional pollution control agencies on interstate rivers. Where the river basin in unmanageably large, such as the Mississippi River system, the EPA might do the planning and take the responsibility itself, in conjunction with the states concerned, letting Congress authorize the water quality standards to be adopted. For areas such as the New York Harbor area, there should be a renewed attempt to create an interstate authority, however, it should be confined to the lower part of the Hudson River and estuary, so that an equal apportionment of state representation between New York and New Jersey, with federal chairmanship, would be realistic. If such an agency could not be created, the EPA itself should take over the planning functions, in collaboration with the states.

One major advantage in having a federal agency conduct the planning is that it assumes the role of planner, rather than director to state and local bodies. In the latter case, there is great room for misunderstanding. Unless the local agency is exceptionally well staffed and vigorous, it may find itself completely dominated by federal influence and budgetary control, while being held responsible in the event that anything goes wrong. On the other hand, able federal officials may have to stand by and see an effort mishandled, and be unable to correct it. It is easy to exercise enough control to stifle initiative without having enough control to assume the direction of the undertaking.

In considering institutional arrangements, we cannot overlook the metropolitan area planning activities of the Corps of Engineers, involving water pollution control. These are excellent stopgap measures, but stopgap only. Moreover, the Corps of Engineers can only implement plans approved by the EPA. In the Merrimack Basin, the Corps had to direct their plans towards the 1985 goal of "no discharge" of pollution whatsoever, a goal that the EPA does not try to implement in its own areawide planning. This constraint artificially directed all planning towards unrealistic goals. Even where constraints are looser and plans sound, the Corps of Engineers still cannot carry them out.

The U.S. Geological Survey has also done some excellent water quality analysis in the Willamette Basin, and is starting similar studies elsewhere. These studies are a model of an intelligent approach and clearly demonstrate how unrealistic the PL 92-500 formulas can be. However, the USGS analysis also suffers from the ultimate handicap of not being the agency responsible for carrying out water pollution control. The longer range solution is not to encourage competition to the EPA among other agencies, but to give it policies and laws which will enable it to accomplish national goals.

Funding of Areawide Planning

The policies of the EPA have delayed the institution of systematic planning for water quality control processes, while great urgency has been given to facilities planning. While this was good short run politics, it has left the EPA in a very awkward situation. The Section 208 plans, if properly carried out, will inevitably reveal that many plans have been approved and funded which are far from optimal. The easiest way for the EPA to avoid criticism on this account would be to fund as few of the 208 plans as possible, start them as late as possible, and try to direct them away from any reconsideration of the relative economy of prior point source treatment decisions. Because of constraints and tight deadlines on real data gathering and analytical work, many of the first generation 208 plans will be ineffective, and the proposed requirement of part nonfederal funding for future 208 programs would tend to discourage further starts. As a first priority, there should be a major federal effort to fund all of the areawide water quality planning as soon as possible, in order to overcome the faulty overall priorities which have so far delayed the planning effort.

Land Use Planning and Control

The control of land use is beginning to be recognized as necessary to preserve the environmental quality of a region, and as essential to safeguard its water quality. Urban and industrial land use contribute to water pollution not only by their contribution to sewage and other point source wastes, but also by contribution to nonpoint sources and urban runoff. This is true for all water bodies. However, land use control is particularly important for watersheds, for lakes, for underground aquifers and for coastal areas.

The quality of the water supplied by watersheds depends largely upon the extent of pollution from the housing and urban growth upstream. Although it is possible to use water taken from extremely polluted sources, the treatment processes become more expensive and less reliable as the water becomes more polluted. Economic development of watersheds can be allowed, provided that point source wastes are highly treated, and special measures are taken to control nonpoint sources. Plans must be made to meet stream quality standards appropriate for the intended use of the water.

Lakes are rather a special problem, because of their high value as recreation areas and for homesites, and also their great vulnerability to eutrophication. Lakes with otherwise pure water can suffer highly adverse eutrophic conditions. As discussed in Chapter 4, the key factor in managing lakes is apt to be control of nutrients, especially phosphorus. Recent studies[7] indicate an increasing interest among the states in preserving lakes in an attractive condition, through control of land use. Measures taken include minimum lot sizes, buildings set back from the water, preservation of wet lands, prohibition or limitation of septic tanks, and control of agriculture or other sediment-producing activities. Prohibiting the use of fertilizer within a given distance from the water's edge would also help. Control of land use is usually exercised within 1000 feet of the lake. Lake protection zoning programs have been adopted by the states of Wisconsin, Minnesota, Vermont, Maine and Washington, and California and

Nevada has a special commission to protect Lake Tahoe. Minnesota has classified its lakes into three categories: (1) natural environmental lakes, (2) recreational development lakes, and (3) general development lakes. Lake shores are zoned for all three categories, with the larger lots and greater setbacks required for category (1), and next, category (2).

Another water resource normally unprotected by zoning is the aquifer. Most publicity concerning aquifers has related to salt water intrusion from the coasts, or salinity due to excessive drawdown. However, aquifers can be adversely affected by general urban development in two other ways. First, the increase in pavements and roofing limits the absorption of precipitation into the ground, thus increasing storm runoff and decreasing the recharge upon which the aquifer depends. Second, pollution from land fills, industrial sites, and many other sources enters into the groundwater, and may make it dangerous to use by reason of bacterial or virus contamination, excess nitrates or toxics.

Wetlands and coastal zones generally are environmentally fragile and must be protected by special zoning regulations. Past practices of leveling off barrier beach sand dunes to provide building sites have proved disastrous and are no longer permitted and speical land use controls on development in coastal zones are now generally required. However, as brought out in Chapter 4, organic material and other nutrients entering estuaries and the sea have entirely different consequences from the same substances entering lakes or impoundments. The environmental consequences of pollution in wetlands are not well understood in detail. Also the relative ecological roles and values of various types of marsh have not been well defined. Therefore there is insufficient scientific basis to guide the land use controls which have been enacted into law. Marshes and swamps in the interior are usually not afforded the same special protection as those of the coast, although they too may shelter important ecosystems.

In general, some degree of control over land use is required in order to assure preservation of water quality, especially in rapidly developing regions. Therefore there must be some relationship between land use control or zoning and water quality planning. However, as yet there is no legislation which makes general land use planning a federal function. It seems likely that, as a result of areawide water quality planning under Section 208, certain limitations in land use necessary for water quality control will be incorporated into the planning of land use by the responsible authorities. (At the present time the responsible authorities are predominantly local state subdivisions.) Since land use control or zoning is so closely related to important functions of local and state government other than water quality control, it does not appear feasible to fully integrate land use planning and water quality planning and use a common approach. Such an approach would imply a monolithic form of government quite different from our present federal system and water resource planners need not yearn after the ideal of an integrated systems approach.

Combined (Comprehensive) Planning Approach

The point has been made numerous times so far that the most advantageous water resource planning must generally incorporate consideration of several different functions or programs, including, in addition to water quality, water supply, land use control, and often also flood control and irrigation. It has also been made clear that although a single integrated approach to these various types of problems would be much simpler and theoretically more satisfactory, such an approach is very unlikely to be adopted in the foreseeable future. This does not prevent our developing combined solutions for the various problems; it does, however, require us to devise the right kind of approach.

In what follows, the term "combined planning" is used to distinguish planning which includes a number of functions rather than only one. That is, it provides for solving problems related to water quality, water supply and sometimes flood control in combination, rather than considering each problem independently. This sort of planning used to be called "comprehensive," but this term has been so overused that it no longer had any precise meaning, and a new term may make the point more clearly.

If we assume that water quality will be evaluated as outlined earlier in this chapter, based upon a least-cost solution to an adopted environmental goal (or alternate goal), the problem then is to combine the water quality planning with water supply and flood control, planning which are based on economic costs and benefits and environmental aspects (as outlined in the "Principles"). The most probable program areas of joint concern in preparing such a combined plan would be: (a) reservoir flow augmentation of rivers, (b) flow reduction of rivers by water supply withdrawals, and (c) runoff detention structures used jointly for flood control and pollution control. To systems analysts, the development of a quantitative optimization process under which the combined sets of conditions would present great theoretical complications; no precise formulation of optimization procedure is possible. However, in practice there is no great difficulty, because we can afford to accept an approximate solution. It is better by far to outline several alternate solutions for a combined plan and to choose the best, rather than to continue with the present approach of developing a water quality plan with most of its provisions established arbitrarily, and then preparing a separate plan for flood control or water supply, with the two plans more apt than not to be mismatched.

There is, of course, no assurance that a combined plan will provide a solution more beneficial than plans prepared for each function considered separately. This is true of all comprehensive planning; the multiple-purpose solution is not always the best. If combined planning reveals this to be the case in any area, nothing will have been lost; the single purpose plans can then be approved and carried out independently.

Of course, the preparation of a combined plan would be much simpler if a single planning agency had complete responsibility for all of the water functions in a given area; this would allow the use of a more closely integrated approach, but it is not absolutely essential. For example, the state of New Jersey is currently preparing a statewide water supply plan, for which it is completely responsible. Concurrently, the county of Middlesex in New Jersey is preparing an areawide water quality plan, under EPA rules and supervision, in which the state is participating. Also, the county is permanently responsible for land use planning, in conjunction with the municipalities concerned. The fact that three different procedures and responsibilities are used is a matter of governmental organization; it prevents adopting a fully integrated approach, but allows coordination, which can achieve the essentials of combined planning. What is essential is that (1) the water quality planning consider and evaluate different levels of stream flow which may result from future water supply withdrawals and storage, (2) the water supply planning consider the possible impacts of future plans on stream flow, and (3) both federal and state policy makers, using comparable data, together decide to what extent the plans for water quality and water supply should be combined or adjusted in the public interest. Similar adjustments must be made between water quality and land use planning. At the present time, there is no requirement that Section 208 planning be coordinated with either water supply or flood control planning to create a combined plan, and there is no established procedure for doing so. The federally financed comprehensive planning processes under the Water Resources Council are supposed to prepare combined plans for all functions, but funding is totally insufficient; and they cannot go far without water quality planning. The Corps of Engineers is required by law to coordinate water supply and flood control planning with other agencies, but there is no similar requirement for the EPA, and, of course, the states may or may not coordinate with federal agencies.

As outlined previously, only a few changes would be required to obtain a combined plan from the present Section 208 approach and the inclusion of water supply planning would be relatively simple. Water quality objectives would be defined first, along with the projected future needs or requirements for water supply. The development and evaluation of programs for achieving water quality would be amplified to include alternative plans which would vary according to expected future stream flows. Similarly, the water supply plans would consider alternative plans and corresponding costs. The scope and cost of each plan would depend upon the extent of utilization of existing stream flows for water supply, and the extent to which reduced stream flow at critical times would be augmented by reservoirs. Therefore, if water supply usage decreased stream flows, and thereby increased the costs of assuring satisfactory water quality, it could be argued that the additional costs should be charged to water supply. However, a case could also be made that water quality control programs requiring minimum stream flows should compensate for the additional costs of obtaining needed water elsewhere. Present legislation provides no means by which either type of compensation could be accomplished. The relative priority of the two programs should be considered by the states and federal government, but above all, each program should be required to consider the interrelationship.

In the case of reservoir flood control, the results of storage on water quality may be either favorable or unfavorable. Large storage reservoirs often have unfavorable impacts on water quality, even to the extent of completely deoxygenating a river. However, since long-term storage is not required for flood control, temporary storage of flood peaks in large reservoirs may have favorable water quality effects by reducing sediment content and the erosive power of the flood waters downstream. The most interesting possibilities of combining water quality control and flood control in the public interest may come through detention of urban runoff. As we have seen in Chapter 4, urban runoff contains large amounts of sediment, organic pollution, heavy metals and petroleum, and it will be necessary to treat much of it if we are serious about controlling the quality of our streams. According to the Staff Report of the National Commission on Water Quality[8] the collection and storage of urban runoff for water quality control might cost $92 billion. In many cases polluted storm sewers and minor tributaries also contribute to flood damage. If such storm runoff could be stored in detention basins for a day or so, until the storm was over, much of the particulate matter could be retained. Since the sediments themselves are environmentally harmful, and most of the heavy metals and petroleum and a good deal of the organic pollution are contained in the sediments, the retention itself and the usual primary treatment procedures would greatly improve water quality. Of course, the solid particulate matter retained would have to be excavated and disposed of periodically to prevent the basins from being clogged. In appropriate cases, a combined approach to flood control and water quality might be considerably less expensive than handling the two problems separately as indicated by recent studies in Chicago.[9]

The present institutional barriers of obtaining approval and funding of both the Corps of Engineers and the EPA appear just about insurmountable. Some means of authorization and financial support of combined plans will have to be found if their potential advantages are to be realized. The problems and the potential savings involved are too great for us to accept two separate plans as the best that can be done.

Under present federal statutes, general flood control is paid for entirely by the federal government, local flood control requires local participation of perhaps fifty percent, water supply storage in federal reservoirs is paid for one hundred percent by local agencies but under financing terms which constitute a major subsidy, and specified pollution control programs have a federal subsidy of seventy-five percent of the capital cost. This diversity in extent of federal participation may not be entirely rational, but that need not unduly concern us. What concerns us more is the restriction of federal support of water quality control to only specified programs of municipal waste treatment and sewer systems. Federal participation should be extended to all types of public water pollution control. If this were done, Chicago's proposed combined water quality control/flood control basins would be eligible for a federal subsidy to assist with the costs allocated to water quality, and the flood control costs (and the construction) could be accomplished by the Corps of Engineers within the framework of their normal authorization and appropriation processes. It is true that considerable coordination would still have to be accomplished between agencies, but this would not be impossible if Congress enacted legislation to support combined planning with appropriate federal assistance.

Illustrative Example: Water Quality Planning for the Delaware Estuary

The River

The Delaware River, and its critical portion, the upper estuary, has been intensively studied for water quality control, however, for a variety of reasons— some of them institutional—the planning for this important region is still far from satisfactory. Moreover, provisions for handling pollution from urban runoff will require a major modification of the approach (see Figure 8-4).

Interagency Plan

The most recently published interagency plan for the river is simply inadequate. The interagency framework plan for the Northeast (North Atlantic Regional Study, 1972) applied arbitrarily selected standards of ninety percent reduction of BOD and ninety-five percent reduction in suspended solids to all basins. The capital expenditures for waste treatment were estimated to total almost $14 billion for the Delaware River, by 2010. However, anticipated industrial growth was so great that it was estimated that net organic waste loading remaining in the river would rise more than sevenfold, which would be catastrophic. The approach is completely unacceptable and the planning was not carried further. If $14 billion is to be spent on the Delaware River for pollution control, with an anticipated environmental disaster at the end, it is economically absurd not to consider what the alternatives might be, but the agencies primarily responsible appear to have suspended their basic planning. A consideration of the relative merits of various control possibilities in the estuary follows.

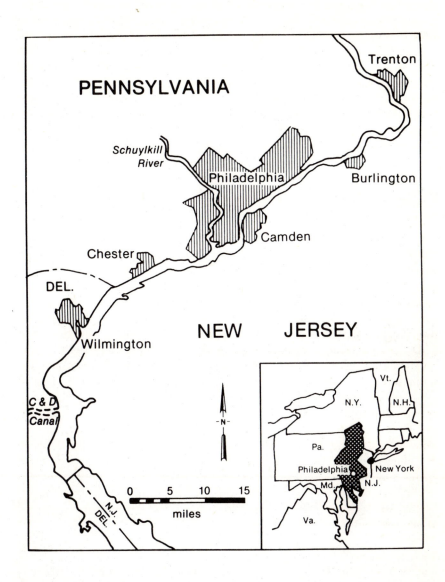

Source: Adapted from the report of the Federal Water Pollution Control Administration. Washington, D.C.: Department of the Interior, 1966.

Figure 8-4. The Upper Delaware Estuary

Earlier Planning

Some of the earlier planning for the Delaware Basin was quite good. The pioneer "preliminary" study of the basin (Federal Water Quality Administration, 1966), outlined alternative programs of water quality control for the estuary, with the costs and effects of each. Although this planning was never officially finalized, certain aspects were continued and brought up to date by the Delaware River Basin Commission staff (Wright and Porges, 1970), taking into account projected growth improvements in flow regimen by planned reservoirs. This study was based mainly upon the dissolved oxygen indicator, which is important to the preservation of the fish life of the river and estuary.

The cheaper of the two alternative plans would cost $34.4 million annually to implement (at then-existing prices) and the other plan $40 million annually. In the summer, the average dissolved oxygen level (DO) in the most critical reach would be 3.5 mg/l under the first plan and under the other 4.5 mg/l. It would cost about $9 million annually for the last 1 mg/l improvement of DO for the estuary. Although it was recognized that any given average level of oxygen would be subject to large deviations, and that downward variations could be harmful to fish, this inherent variability was merely treated statistically at the time and was used to justify establishing the minimum levels above 3.5 mg/l.

The Delaware River Basin Commission staff realized that all effluents on the river should not be treated alike. Removal of an average of eighty-five percent of BOD was considered necessary, but not in all zones, and much greater economy could be obtained by differential treatment, depending upon the location of the waste source in question. Such an approach employs principles familiar to resource economists. Minimum total costs of a program can be obtained by optimization between treatment and other alternatives such as process changes, and by taking advantage of marginal cost and marginal effect differentials. However, all such approaches are now somewhat questionable in the Delaware and elsewhere because of the uniformity of treatment which appears to be intended by the 1972 Act.

Maintenance of Flows

The maintenance of satisfactory minimum flows in the Delaware River is also a matter of concern. The high level of anticipated municipal and industrial uses of the basin's fresh water resources will impose a severe stress on the area's high quality natural environment. A minimum river flow of 3000 cubic feet per second at Trenton has been adopted for planning purposes, but the planned maintenance of this flow during dry periods was based upon completion of the authorized Tocks Island Dam. This project encountered prolonged environmental objections and is now not expected to be built. Moreover, it has been found that enormous electric energy demands of the future will require considerably more water for evaporative cooling, than had been allowed for in the planning. It is difficult to obtain official recognition of the importance of flow maintenance to water quality, because of the EPA doctrine that treatment as specified in the 1972 Act is an unqualified absolute, unrelated to any other considerations. Any sensible planning for water quality in the Delaware Estuary must include a plan for maintenance of satisfactory low flows, or alternatively of coping with the deficiencies in flow. This may seem obvious to engineers and analysts, but it seems that EPA planning cannot or at least does not consider such aspects.

Urban Runoff as a Pollution Source

The new insights into unrecorded and nonpoint sources of pollution from urban runoff have added a new consideration to planning. Previously, allowances for pollution from urban runoff were made on an arbitrary basis, treated as a matter of minor importance, or neglected altogether. Recent studies have provided a basis for approximately estimating the organic pollution from urban runoff. The urban areas on the upper Delaware Estuary from Trenton, New Jersey to Wilmington, Delaware, inclusive, have a population of 2,992,000 (1970 census), which might produce an annual average of 74,800 lbs. of BOD daily throughout the year.[10] The greatest amount of urban runoff takes place following storms, which occur only periodically. A single severe storm can cause an excess in average flow for seven days, or perhaps 525,000 lbs. of BOD.

There are other ways to estimate the magnitude of the urban runoff loading. For example, the total area occupied by the estuary urban population is 287 square miles. Using the STORMmodel coefficients and available land classification data, an annual runoff of twenty-eight inches can be computed. This means that from a two-inch general rainfall, if the BOD were 10 mg/l, the total urban runoff loading would be 53,000 lbs. Mean values of 10 mg/l of BOD in urban runoff during storms seem to be reasonable, as far as can be determined from the literature. Research is under way to obtain better estimates.

There is a third way in which the urban runoff pollution reaching the Delaware Estuary can be estimated, and that is through the levels of dissolved oxygen in the estuary following a storm. Analysis of Delaware Estuary DO in critical areas indicated that the decrease in DO after a storm averaged 2.05 mg/l.[11] The estuary took an average of six days to reach the minimum DO following a storm, plus another two to six days to recover. This confirms the massive organic shock loads of approximately one-half million pounds computed as described above. Since the commission hopes to reduce total loading of oxygen demand to about 215,000 pounds of BOD daily, it is apparent that urban runoff will greatly influence the water quality.

Control of Pollution From Urban Runoff

As indicated in Chapter 4, there are a variety of ways to control pollution from urban runoff, all of which are necessarily expensive because of the extreme variability of occurrence. The cheapest way to control major proportions of BOD and sediments from urban runoff appears to be short term detention followed by primary treatment, which might cost something like forty cents per pound of BOD removal during storm events. BOD in levels characteristic of the Delaware is not harmful in itself; the main objective of lowering it is to maintain oxygen levels in the receiving waters. The dissolved oxygen levels may be maintained in large rivers by other means on a much more economical basis, as indicated below.

Studies and field tests of artificial aeration on the Delaware Estuary previously referred to have shown that with a continuous summer operation, the last milligram per liter of dissolved oxygen could be added to the river by mechanical and diffuser aerators at about one quarter the incremental costs of waste treatment to achieve the same end. Furthermore, subsequent research showed that, a mobile oxygen dispersing craft could reduce costs still further. This study showed that a mobile oxygenating craft dispersing 400,000 pounds of oxygen daily could be built and operated on the Delaware Estuary for total costs, including amortization, of about $1,300,000 annually. For urban runoff, the craft would only be operated about one-fourth of the time, at most, which would reduce total costs by about thirty-five percent. If transfer efficiency were seventy-five percent, and the craft operated thirty days of a five-month period, 9,000,000 pounds of oxygen would be imparted to the river and unit costs would be about nine cents per pound, based upon 1972 costs. This is probably the cheapest way of adding additional oxygen to a large navigable river.

Of course, BOD is not the only pollutant involved in urban runoff; there are usually considerable loadings of heavy metals and petroleum, which are mostly associated with sediments. The detention and primary treatment of urban runoff would also remove these other pollutants, and therefore might be desirable in some of the tributary basins even though the benefits to the oxygen level through reducing BOD alone would not be worth the costs. The detention and treatment of the "first flush" of storm runoff would be particularly valuable from a water quality viewpoint.

An Environmental Planning Approach

Water quality planning for the Delaware Estuary or any other major waterway should start with an environmental objective. The environmental objective has been set for dissolved oxygen for the Delaware and with further research it could be established for other pollutants as well. For example, a major study will be required to determine whether nutrients are harmful or not in the lower estuary. Also, social goals are relevant.

The plan for water quality control should be based upon a projection of future pollution conditions, assuming at least secondary treatment of all wastes; analysis of origins of pollution including urban runoff and other unrecorded wastes; and a projection of future water supply consumption and of diversions from the basin.

Possibilities of storage for low flow maintenance should be considered in the plan, so that costs of maintaining or improving low flows after depletion by upstream diversion could be estimated, recognizing that the Tocks Island Reservoir will not be available.

The study of pollution origins should not only consider BOD and dissolved oxygen, but also sediment, nutrients, and toxic substances, such as heavy metals and biocides.

The various possibilities of remedial measures should be combined into alternative plans, to determine the most economical way to achieve given environmental objectives. The DRBC system of allocation of a given loading to a facility would be used in preference to specifying an effluent standard by concentration.

The plans should include consideration of all alternatives which may contribute to the given objective. The final plan for the estuary might include the following elements:

A. Secondary treatment of all organic wastes, including ammonia oxidation at large plants, with the percent of treatment varying from zone to zone
B. Supplemental treatment to limit acids and various toxics in specific cases
C. Oxygenation of the river all summer, with supplemental oxygen for several days after storms
D. Special controls on sediment where construction is underway
E. Treatment of some "first flush" urban runoff
F. An extensive program to locate and control leaky sewer systems, poorly regulated combined sewers, unauthorized and unrecorded commercial and industrial wastes, and other unrecorded sources
G. Certain provisions of land use control
H. No requirement to remove phosphates or nitrates
I. A halt in chlorination of effluents and substitution of other means of disinfection, particularly at Trenton
J. A plan for low flow augmentation combined with flood control

Relationship to Present Laws and Institutions

The program suggested above for the Delaware River is a forecast of what might be developed by a real combined plan. It follows basic principles of environmental objectives and resource economics, but may or may not be consistent with the provisions of PL 92-500. The law, as it is, does not require anything like the suggested combined approach, nor does it appear to allow any major deviation from national standards except for tightening them in particular cases.

Implementation of Plans Under a New Approach

The nation is thoroughly involved in the present program of water pollution control, which has been criticized, but at least has the merit of considerable momentum. Obviously, we could not afford to bring the entire program to a halt for several years until final plans could be completed for each area. On the other hand, it would not be wise to allow the EPA to proceed with its present procedures until new types of combined plans could be prepared, but the extreme slowness of the EPA and its predecessors suggest that, if this approach were adopted, the new plans would never be finished. Instead, it would be desirable to set up as a prerequisite to decision making, a system of interim plans, which could be issued by the EPA, but which would be sent to Congress for approval in the event the state concerned did not agree with them. This process would work as follows:

A. For situations where the EPA or a state has issued approvals or orders involving advanced waste treatment under authority of PL 92-500, and the industry or municipality feels that the requirement is unreasonable and will not be sustained by combined water quality planning, it should be allowed to temporarily withhold installation of advanced waste treatment until the completion and approval of another plan; but, if that plan sustains the requirement for advanced treatment, it should then be implemented promptly. (This procedure should not be used to limit removal of toxic substances where public health is involved.)

B. Pending approval of final combined plans, authority should be given to the EPA to approve interim areawide or basin plans prepared by any agency, or to issue interim EPA plans, based upon the principles of combined planning; plans should be subject to approval by Congress if the state concerned disagrees with them, but otherwise they should be immediately effective.

C. The EPA should have the power to approve plans prepared by a state or interstate agency or to send them back for reconsideration. However, if there remains a difference of opinion between the EPA and the state or interstate agency, the EPA should write a review report to Congress, explaining the plan's deficiencies and recommending a revised plan. Upon approval by the Congress the revised plan should be put into effect.

D. When an interim or final plan has been approved, and required implementing legislation has been passed by the state, the state should be allowed to initiate the individual projects promptly with block federal funding. If the state fails to act promptly, if the measures proposed to be executed first are not the most effective of the measures outlined by the plan, or if the measures taken are ineffective, the EPA should be authorized to suspend the execution of the plan, and, after carrying out hearings, make a report to Congress recommending appropriate modifications of the plan or of the implementing program.

E. Funding should be provided to accelerate the preparation of final combined plans nationwide, to supersede the interim plans as soon as practicable. A period of about four years should be allowed to initiate all such plans, and about three years to complete each one.

General Conclusions

The treatment of this book is complex, ranging far afield, and is full of technical explanations of water pollution relationships. This complexity is unavoidable because the interrelationships concerning the chemistry of polluting substances, the regimen of streams, and the ecosystems and man's environment, generally control the social value of various alternative planning approaches. Our waterway systems and the economic activities affecting them are very diverse, therefore, the best means to remedy undesirable conditions are also diverse. The simplicity of wording of PL 92-500 may have been good politics in 1972, and it certainly provides ease and a good legal basis for administration. The disadvantage is that uniform approaches are often not the best solution; they may cost us too much in money, manpower, materials and energy which could be better used elsewhere.

Protection of the environment cannot be considered by itself as it is in competition with many other socially important public objectives, including public health, education, a high rate of employment, and help and service to the disadvantaged. The water quality program is not going to have available in the near future the immense sum of 511 billion dollars estimated as necessary to meet in full the 1983 objectives,[12] and it had better find approaches to make sure that low cost alternatives and high priority goals are selected first. It is dismaying to contemplate the continued environmental pollution of many streams despite the treatment programs: the fact that the Delaware River still has DO levels falling below 1.0 mg/l in summer, the fact that gross pollution still oozes, seeps, or is dumped in many unrecorded urban and industrial wastes, the fact that most of our reservoirs and lakes in populated areas are scummy and at least fairly odoriferous in summer, and the fact that much of the damage to ecosystems comes from pollutants and sources not yet identified, measured and programmed for control. This book is intended to reveal as clearly as possible the obscure channels by which pollution works its environmental damage, and the lack of precision and adequacy of our present governmental approaches. We ought to be able to do much better than this, and with more knowledge.

In spite of very large sums which have been expended, there is a great lack of data as to our river and estuarine ecosystems, and the impacts of pollutants upon the receiving waters and their biota. Much more research is required. We also lack sufficiently developed planning technology for systems analysis to obtain optimum water pollution control and combined plans, considering different functions.

We need better government organization to facilitate the planning and management of complex regional systems, however, we need not wait for government adoption of ideal types of organization or planning approaches. The biggest advantages of planning more effective systems can be obtained by coordination rather than integration, while using different planning staffs and different criteria. It is essential that Congress make it clear that coordination and combined planning are seriously meant, and establish federal cost-sharing to be made applicable to any of the means found optimum to obtain water quality objectives and not just certain ones. No matter what changes may be made in our planning institutions, including differences in various regions depending upon state law, the types of approach outlined in this book will result in more environmental value for the resources expended. We cannot afford to indefinitely accept the delays and inadequacies of planning for water resources development and water quality control under the present systems.

Appendixes

Appendixes

Appendix A
Recommendations of the
Ad Hoc Committee

1. The national objective of water quality planning should be the achievement of environmental quality and related social goals at minimum costs in terms of human and economic resources or, as stated by the U.S. Water Resources Council, the twin objectives of economic efficiency and environmental quality.

2. Regional goals of water quality planning should be specified for each state by the state concerned or for an interstate region or river basin by its legally constituted interstate planning agency, if one exists, subject in either case to approval by the EPA, and to specific authorization by Congress. The goals specified should be stream quality standards for each basin reconsidered on the basis of knowledge acquired in the last ten years, to be maintained with a given degree of reliability for a given regimen of stream flow, and considering other related environmental objectives, including suitability for habitat.

3. The selection of means to be adopted for water quality control should be made as the result of analysis and planning, considering the actual circumstances of each area or basin, and not by application nationwide of rules or legal definitions.

4. The national water quality control program should relate to pollution from all sources, and take advantage of a variety of alternative approaches, including nonstructural measures, rather than being confined to treatment of effluent wastewater.

5. Water quality control plans should be comprehensive; they should make provision for and take fully into account the future water supply needs and withdrawals for regions concerned, and must also recognize and make provision for appropriate future usage of waterways and adjacent areas as wildlife habitat and for recreation.

6. Water quality control plans should be developed so as to achieve the maximum environmental benefit at least cost, no matter whether the private or the public sector of society is responsible for a given alternative.

7. The federal financing of pollution control improvements should be on a basis such as to encourage adoption of the least-cost alternative to obtain a given objective. Therefore, within the category of governmental action programs, federal financing should support a given proportion of the costs of all alternatives, including not only treatment of wastes, but also any relevant storage programs, instream improvements or control of urban runoff. Similarly, industry should be required to reduce its discharge of polluting wastes without setting up incentives which would tend towards adoption of less than optimal programs. This can be done by alloting industrial effluent loadings, thus allowing

industry the option of accomplishing the reduction by treatment, by limitation of output, or by process changes.

8. The nonpoint sources of pollution in rural areas should be evaluated and, where necessary, corrected by such means as sediment and nutrient limitations, land use zoning, and control of solid waste disposal. These controls can be enforced and monitored by governmental agencies.

9. Pollution from urban and suburban runoff is a major problem in many areas: (1) pollution from industrial and commercial runoff, leakages and illegal connections should be controlled by the owners under direct government supervision; (2) the municipality itself should take responsibility for programs of controlling or treating leakage from city sewer systems, storm and combined sewer discharges; (3) other minor tributary and sheet flow runoff should be financed as an alternative to point source waste treatment under provisions of paragraph 7 above.

Names of the committee members are as follows:

William Whipple, Jr., Chairman; Rutgers University

Bernard B. Berger; University of Massachusetts

Charles P. Gates; Cornell University

Warren A. Hall (by letter); Colorado State University

Joseph V. Hunter; Rutgers University

Ruben Johnson; Consultant

Forest C. Neil; Greater Chicago Metropolitan Sanitary District

Joseph V. Radziul; Philadelphia Water Department

Robert M. Ragan; University of Maryland

Clifford W. Randall; Virginia Polytechnic Institute and State University

Clarence J. Velz; Consultant

William R. Walker; Virginia Polytechnic Institute and State University

James C. Warman; Auburn University

Appendix B
Technical Note

In order to avoid objections in principle to adding quantitative utilities of different persons, a theory of economic optimization was developed by Vilfredo Pareto, 1848-1923, an Italian economist and sociologist. Efficiency in the Pareto sense means that the distribution of consumer goods should be such that any further adjustment in allocation to consumers would result in the reduction of satisfaction of at least one person. Production is Pareto optimal if every input is allocated to firms so that any change will result in the reduction of output of at least one firm. This seemingly complicated method for specifying efficiency stems from the desire to avoid aggregating individual utilities to achieve a social welfare function. The search for Pareto optimality can only be guided by the principle that any change which makes at least one individual better off, and none worse off, is an improvement in social welfare.[1] The Pareto approach was developed into an elaborate system of economic theory, designed to avoid interpersonal comparisons of welfare or utility, by Kaldor,[2] Scitovsky,[3] Little[4] and others. However, Baumol[5] definitively shows the sterility of such approaches when applied to practical problems. Briefly, the Pareto optimality approach does not provide guidance for any situation in which a proposed action would benefit some persons to the disadvantage of others. With regard to taxes, almost every problem of public policy in environmental situations involves conflicting interests between concerned groups, and there are various other objections to the other criteria proposed.[6] Although intellectually useful in approaching certain theoretical problems, particularly through the medium of a technique known as individual indifference curves, the welfare theory approach has extremely limited possibilities. Curiously enough, many economists feel it appropriate to refer approvingly to Pareto optimality as the basis for their studies, even though in fact they aggregate the utility of various persons by techniques such as preparing community indifference curves. If utilities of different purposes are to be aggregated in the end, the entire basis for the theoretical welfare economics approach is nullified. It is better for practical purposes never to start on such an approach.

Of course, the basic position which motivated Pareto in the first place remains: there *are* differences in utilities of a given economic unit to different persons. We may assume that there are quantitative differences in the marginal utility of monetary values to individuals, although these differences are unknown and unknowable. However, the mean differences between individuals' utilities become insignificant in their effect when we are comparing the aggregate net utility or disutility of some proposal to large groups of people, who may be assumed to be randomly selected. According to basic probability theory, if the mean deviation of any normally distributed stochastic variable for a single

205

person is X, the mean deviation for a group of n persons is $X/\sqrt{n}$. Thus, for a group of 1000 persons, the mean deviation from normal of the aggregate of all these increments to utility would be only about three percent of the deviation most probably to be expected of a single person. Therefore, when dealing with large groups of individuals, individual differences in marginal utility of money are relatively unimportant, and only class differences are of practical concern.

Notes

Notes

Chapter 1
Water Pollution, A National Problem Area

1. Warren D. Fairchilds, "Necessity for Planning," in "Urbanization and Water Quality Control," William Whipple, Jr., ed., American Water Resources Association, 1975.

2. *Water Newsletter*, vol. 17, no. 5 (March 1975).

3. Environmental Protection Agency "National Water Quality Inventory: 1975 Report to Congress." EPA-440/9-75-014. 1976.

4. Morris A. Wiley, "The Petroleum Industry and Cost-Effective Water Quality Planning," in *Urbanization and Water Quality Control*, edited by William Whipple, Jr. (hereafter cited as *Urbanization*) (Minneapolis: American Water Resources Assoc., 1975) hereafter cited as Wiley, "The Petroleum Industry and Water Planning."

5. *Certain Recommendations of the Water Pollution Control Federation for Improving the Law and Its Administration* (Washington, D.C.: Water Pollution Control Federation, 1975).

6. Ibid.

7. Ibid.

8. "Water Pollution Control Federation Looks at Water Cleanup Problems and Prospects," *Environmental Science and Technology* 8:1073 (December 1975).

9. "Research and the Quest for Clean Water," *Journal of the Water Pollution Control Federation*, vol. 47, no. 2 (February 1975):240-251.

10. Ibid.

11. General Accounting Office, *Research and Demonstration Program to Achieve Water Quality Goals: What the Federal Government Needs to Do*, B-166505 (Washington, D.C.: Government Printing Office, 16 January 1974).

12. National Utility Contractors Association, *Fulfilling a Promise–An Analysis of the 1972 Clean Water Act's Construction Grants Program* (Washington, D.C.: National Utility Contractors Association, 1975) hereafter cited as Utility Contractors, *Fulfilling a Promise*.

13. J.T. Ling, "The High Cost of Getting Water Too Clean," *Wall Street Journal* (1 August 1972).

14. *Implementing the National Water Pollution Control Permit Program: Progress and Problems*, Comptroller General of the U.S. (9 February 1976).

15. Ibid.

16. Enviro Control Inc., *Total Urban Water Pollution: The Impact of Stormwater*, report for the Council on Environmental Quality (Washington, D.C.: Council on Environmental Quality, 1974), PB 231-730.

17. *Staff Draft Report*, National Commission on Water Quality (November 1975) hereafter cited as *Report*, Commission on Quality.

18. Russell W. Peterson, luncheon address at the Annual Meeting of the Water Pollution Control Federation, 1974.

19. Charles F. Luce, before the Finance Club of the Graduate School of Business Administration, Harvard University, Cambridge, Mass., 9 December 1975.

20. A.V. Kneese and B.T. Bower, *Managing Water Quality: Economics Technology, Institutions* (Baltimore: Johns Hopkins Press, 1968) hereafter cited as Kneese and Bower, *Managing Water Quality*.

21. William Whipple, Jr., "Water Quality Planning for the Delaware Estuary—An Example," Water Resources Bulletin, Vol. 11, No. 2, April 1975.

22. W. Whipple, Jr., "BOD Mass Balance and Water Quality Standards," *Water Resources Research* 6 (1970):827, hereafter cited as Whipple, "BOD Mass Balance."

23. *Report*, Commission on Quality.

24. G.E. Wood, "Midcourse Correction," remarks at a Water Pollution Control Federation in Washington, D.C., April 1976.

25. "Report to the Congress by the National Commission on Water Quality," National Commission on Water Quality, 1976, hereafter cited as "Report to Congress," Commission on Water Quality.

Chapter 2
The Environmental Quality Objective

1. C.R. Goldman et al., *Environmental Quality and Development*, 2 vols. (Davis: University of California, 1971). Contract for National Water Commission, NWC-EES-72-032. Hereafter cited as Goldman, *Quality and Development*.

2. F.W. Stearns and Tom Montag, *The Urban Ecosystem* (Stroudsburg, Pa.: Dowden, Hutchinson & Ross, 1974) hereafter cited as Stearns and Montag, *Urban Ecosystem*.

3. L. Gerlach, "Mobilization and Participation by Citizens Groups in Improving Quality of Water Resources Environment," *Water Resources Research Center Bulletin* 57 (Minneapolis: University of Minnesota Press, 1973).

4. "Encounters with the Archdruid," *New Yorker*, 20 March 1971, 27 March 1971, 3 April 1971.

5. I. Burton, "The Quality of the Environment, A Review," *The Geographical Review* 58 (1968):3.

6. H. Meadows Donella, "The Limits to Growth," report for the Club of Rome's Project on the Predicament of Mankind (New York: Universe Books, 1972).

7. R. Dubos, *So Human An Animal* (New York: Scribner, 1968) hereafter cited as Dubos, *Human Animal.*

8. Ibid.

9. Goldman, *Quality and Development.*

10. W.A. Thomas, *Indicators of Environmental Quality* (New York: Plenum Press, 1972).

11. Goldman, *Quality and Development.*

12. "Beauty of America," proceedings of the White House Conference on Natural Beauty, 1965.

13. T.H. Hall, *The Silent Language* (New York: Doubleday, 1969).

14. T.H. Hall, *The Hidden Dimension* (New York: Doubleday, 1966).

15. Paul L. Errington, *Of Men and Marshes* (Ames: Iowa State University Press, 1957).

16. Hoch, Irving, "Urban Scale and Environmental Quality," Resources for the Future, Inc., Reprint 110, August 1973.

17. O.A. Salama, *Planning and Human Values* (Cambridge, Mass.: Abt Associates, April 1974).

18. Stearns and Montag, *Urban Ecosystem.*

19. Rachel Carson, *Silent Spring* (Boston: Houghton Mifflin, 1962).

20. T.G. Bahr, et al., *Recycling and Ecosystem Response*, The Institute of Water Research (East Lansing: Michigan State University, February 1972) for the National Water Commission.

21. Ibid.

22. R.S. Bierman, "Science and Value," *Science* 178:40-60, 3 November 1972.

23. S.V. Ciriacy-Wantrup, *Water Policy and Economic Optimizing: Some Conceptual Problems in Water Research*, American Economic Review, vol. 57, no. 2, May 1967, pp. 179-189.

24. Dubos, *Human Animal.*

25. R. Dubos, *Man Adapting* (New Haven: Yale University Press, 1965).

26. E. Neumann, *The Great Mother: An Analysis of the Archetype* (Pantheon Books, 1955) hereafter cited as Neumann, *The Great Mother.*

27. Dubos, *Man Adapting.*

28. Genesis 1:28.

29. Goldman, *Quality and Development.*

30. Ibid.

31. L. White, Jr., "The Historical Roots of Our Ecological Crisis," *Science* 155 (1967):1203.

32. L. Marx, "American Institutions and Ecological Ideals," *Science* 170 (1970):945.

33. G. Santayana, *A Sense of Beauty* (New York: Dover Publications, Inc., 1955).

34. M.H. Krieger, "What's Wrong with Plastic Trees?" *Science* 179:446.

35. M.I. Goldman, ed., "Ecology and Economics: Controlling Pollution in the 70's," Englewood Cliffs, p. 108, 1972.

36. L.B. Leopold, "The Conservation Attitude, Part C," *Geological Survey Circular 414* (Washington, D.C.: Conservation and Water Management, 1960).

37. K.E. Boulding, in *Environmental Qualityin a Growing Economy*, by H. Jarrett (Baltimore: Johns Hopkins Press, 1966). Hereafter cited as Boulding, *Environmental Quality*.

38. M. Clawson, *Methods of Measuring the Demand for and Value of Outdoor Recreation* (Washington, D.C.: Resources for the Future, 1959).

39. N.H. Coomber, and Biswas, A.K., "Evaluation of Environmental Intangibles," *Environment*, Canada, June 1972.

40. J.L. Knetsch, "Communications: Assessing the Demand for Outdoor Recreation," *Journal of Leisure Research* 1 (1969):86. Hereafter cited as Knetsch, "Assessing the Demand."

41. J.J. Seneca and C.J. Cichetti, "User Response in Outdoor Recreation: A Production Analysis," *Journal of Leisure Research* (1969):238. Hereafter cited as Seneca and Cichetti, "User Response."

42. J.L. Knetsch and R.K. David, in *Economics of the Environment*, edited by R. Dorfman (New York: Norton, 1972), p. 389.

43. R.P. Mack and S. Meyers, in *Measuring Benefits of Government Investments*, edited by R. Dorfman (Washington, D.C.: Brookings Institution, 1965), p. 83.

44. S.D. Reiling, K.C. Gibbs and H.H. Stoevener, *Economics Benefits from an Improvement of Water Quality*, EPA-R5-73-008 (Washington, D.C., January 1973).

45. M. Clawson, "Natural Resources and International Development" (Baltimore: Johns Hopkins Press, 1964). Hereafter cited as Natural Resources.

46. R.T. Smith and N.J. Kavanaugh, "The Measurement of Benefits of Trout Fishing: Preliminary Results of a Study at Grafton Water, Great Ouse Water Authority, Huntingdonshire," *Journal of Leisure Research* 1 (1969):318.

47. N.H. Coombs and A.K. Biswas, "Evaluation of Environmental Intangibles," *Environment*, Canada, June 1972.

48. *Methodology to Evaluate Socio-Economic Benefits of Urban Water Resources* (East Orange, N.J.: Louis Berger, Inc., 1971).

49. Ibid.

50. C.D. Foster and M.D. Beesley, in *Readings in Welfare Economics*, by K.J. Arrow and T. Scitorsky (Homewood, Ill.: Irwin, 1969), p. 475.

51. Boulding, *Environmental Quality.*

52. Foster and Beesley, in *Readings.*

53. Goldman, *Quality and Development.*

54. Ibid.

55. N.H. Coomber and A.K. Biswas, "Evaluation of Environmental Intangibles," *Environment* (Toronto, Ontario: University of Toronto, June 1972).

56. Knetsch, "Assessing the Demand."

57. Ibid.

58. Seneca and Cichetti, "User Response."

59. U.S., Congress, Senate, *The Policies, Standards, and Procedures in the Formulation, Evaluation and Review of Plans for Use and Development of Water and Related Land Resources*, S.D. 97, 87th Congress, 2d Session.

60. Knetsch and David, *Economics of Environment.*

61. P. Bohm and A.V. Kneese, *The Economics of Environment* (London: Macmillan, 1971), p. 94.

62. Dornbusch & Co., Inc., "A Generic Method to Forecast Benefits from Urban Water Resource Improvement Project," *Office of Water Research and Technology*, November 1974.

63. I. Burton, "The Quality of the Environment, a Review," *The Geographical Review*, 58, 3, 471-481, 1968.

64. D. Zwick and M. Benstock, *Water Wasteland* (New York: Grossman Publishers, 1972) hereafter cited as Zwick and Benstock, *Water Wasteland.*

65. Yu-Fu Tuan, "Our Treatment of the Environment in Ideal and in Actuality," *American Scientist* 58 (1970):244-249.

66. Goldman, *Quality and Development.*

67. H. Osborne, *Theory of Beauty* (London: Routledge and Kegan, 1952) hereafter cited as Osborne, *Theory of Beauty.*

68. Cive Bell, *Art* (New York: Capricorn Books, 1958).

69. Osborne, *Theory of Beauty.*

70. George Santayana, "Reason in Art," *The Life of Reason*, vol. 4 (New York: Scribners, 1922).

71. George Santayana, "The Last Puritan," (London: Constable, 1935).

72. George Santayana, "Reason in Art," in *The Life of Reason*, vol. 4 (New York: Scribners, 1922).

73. Osborne, *Theory of Beauty.*

74. National Water Commission, *An Aesthetic Overview of the Role of*

Water in the Landscape, by R.B. Litton, Jr., et al., NWL-EES-72035, 1971. Republished as *Water and Landscape* (Port Washington, N.Y.: Water Information Center, Inc., 1974) hereafter cited as Water Commission, *Aesthetic Overview*.

75. John A. Dearniger, *Measuring the Intangible Values of Natural Streams, Part II* (Lexington, Ky.: University of Kentucky, 1973) for the Water Resources Research Institute.

76. J.S. Larson et al., *A Guide to Important Characteristics and Values of Freshwater Wetland in the Northeast* (Amherst: University of Massachusetts, July 1973).

77. C.A. Gunn et al., *Development of Criteria for Evaluating Urban River Settings for Tourism–Recreation Use* (College Station, Tex.: Texas Water Resources Institute, June 1974) hereafter cited as Gunn, *Development of Criteria.*

78. Water Commission, *Aesthetic Overview.*

79. Kneese and Bower, *Managing Water Quality.*

80. D.H. Harris, *Assessment of Turbidity, Color and Odor in Water* (Santa Barbara, Calif.: Anacapa Sciences, Ltd., 1975) for the Office of Water Resources Research.

81. *Report*, Commission on Water Quality.

82. Zwick and Benstock, *Water Wasteland.*

83. G. Hill, "Impure Tap Water," *New York Times*, 13 May 1973.

84. L.W. Bone and D.A. Becker, *Pathogenic Free-Living Amoeba in Arkansas Recreational Waters*, Water Resources Research Center (Fayettville: University of Arkansas Press, 1975).

85. B.J. Mehalas et al., *Water Quality Criteria Data Book–Volume 4*, 18040DAZ 04/72 (Washington, D.C.: Envirogenics Co., April 1972) for the U.S. Environmental Protection Agency.

86. Ibid.

87. D.A. Chant and A.H. Clouter, "Parameters for Ecosystem Evaluation," *Environment*, Occasional Paper #1 (Toronto, Ontario: University of Toronto, 1974).

88. I. Burton. "The Quality of the Environment, a Review," *The Geographical Review* 58, 3, 471-81, 1968.

89. "Beauty for America," Proceedings of the White House Conference on Natural Beauty, 1965. Government Printing Office, Washington, D.C.

Chapter 3
Effects of Pollution on Receiving Waters

1. Federal Task Force on an Inadvertent Modification of the Stratosphere, "Fluorocarbons and the Environment," June 1975.

2. J.C. Davies et al., "Decision Making for Regulating Chemicals in the Environment," National Academy of Sciences, 1975.

3. Ronald G. Toms, "Regional Water Quality Control in Great Britain," in *Urbanization*, hereafter cited as Toms, "Water Control in Great Britain."

4. W.S. Broecker, "Environment Priorities," *Science* 182-4111 (2 November 1973).

5. C.B. Stalnaker, and J.L. Arnette, *Methodologies for the Determination of Stream Resource Flow Requirements: An Assessment* (Logan: Utah State University, 1976) U.S. Fish and Wildlife Service.

6. W.J. Oswald and R. Ramaui, "The Fate of Algae in Receiving Waters," in *Ponds as a Wastewater Treatment Alternative*, edited by E.F. Gloyna et al. (Austin, Texas: University of Texas, 1976).

7. Informal communication from Dr. J.V. Hunter, Rutgers University, to the author.

8. Ruth Patrick, Academy of Natural Sciences of Philadelphia, in *Urbanization*.

9. F.A. DiGiano, R.A. Coler, R.C. Dahiya and B.B. Berger, "A Projection of Pollutional Effects of Urban Runoff in the Green River, Mass.," in *Urbanization*, hereafter cited as DiGiano et al., "Projection of Pollutional Effects."

10. U.S. Environmental Protection Agency, *Oil Spills and Spills of Hazardous Materials* (Washington, D.C.: Office of Air and Waste Programs, Division of Oil and Hazardous Materials.

11. Ciriacy-Wantrup, *Conceptual Problems.*

12. S.D. Faust and E.W. Mikulewicz, "Factors Influencing the Condensation of 4-aminoantipyrene with Derivatives of Hydroxybenzene," *Water Research* 1 (1967): 509.

13. C.M. Fetterolf, "Proceedings of Industrial Waste Conference," Lafayette, Ind.: Purdue University 115 (1964): 174.

14. Russell W. Peterson, Luncheon address at annual meeting Water Pollution Control Federation, 1974.

15. G.D. Khamuev, *Gigiena i Sanit.* 32(1-3)15, (1967) Moscow.

16. S.P. Varshavaskaya, *Gigiena i Sanit.* (33(10), 15 (1968) Moscow.

17. L.N. Gurfein and Z.K. Pavolova, in *USSR Literature on Water Pollution Control*, edited by B.S. Levine 3 (1962): 58 (Washington, D.C.: U.S. Department of Commerce).

18. S.G. Davydova, *Girgiena i Sanit.* 32(8), 7 (1967) Moscow.

19. V.N. Fomenko, *Girgiena i Sanit.* 30(1-3), 8 (1965) Moscow.

20. S.Y. Marshtein and E.V. Lisovskaya, *Girgiena i Sanit.* 30(1-3), 177 (1965) Moscow.

21. V.N. Tugarinova et al., *Sanit. Okhrana Vodoemov at Zagryazneniva Prom. Stochnymi Vodami* No. 5 (1962): 285. Teknichescogo Soveshchamya, Kiev.

22. A.V. Sedov, in *USSR Literature on Water Pollution Control*, edited by B.S. Levine, 8, 43 (1966), (Washington, D.C.: U.S. Department of Commerce).

23. V. Zitko, *Bulletin Environmental Contamination Toxicology*, 6(5), 464 (1971), Springer Verlag, N.Y.

24. H.C. Davis and H. Hindu, *U.S. Fish and Wildlife Service Fishery Bulletin*, 67(2), 393 (1967).

25. P.W. Webb and J.R. Webb, *Journal of the Fisheries Research Board Canada* 30 (1973): 499.

26. E.L. Morgan et al., *Virginia Journal of Science*, Hopewell, Va. 23 (1972): 114.

27. Hill, "Impure Tap Water."

28. "Chlorine Cancer Link Study by EPA," *Home News*, 2 November 1974, p. 3.

29. E.J. Genetelli, S.A. Lubetkin and J. Cirello, "Chlorination of Waste-water Effluents, A Review," unpublished state-of-the-art summary for the Passaic Valley Sewage Commission.

30. C.E. Zobell, *Proceeding Experimental Biological Medicine* 34 (1936): 113-116.

31. Larry A. Esvelt, Warren J. Kaugman, and Robert E. Selleck, "Toxicity of Treated Municipal Wastewater," *Journal of the Water Pollution Control Federation* 45 (1973):1558-1572. Hereafter cited as Esvelt et al., "Toxicity of Wastewater."

32. Shu-fa Tsai, *Water Quality Criteria to Protect the Fish Population Directly Below Sewage Outfalls* (College Park, Md.: University of Maryland Press, August 1971).

33. Thomas Bott, Academy of Natural Sciences of Philadelphia, to the author.

34. Ruth Patrick, "Some Thoughts Concerning Correct Measurement of Water Quality," in *Urbanization*, hereafter cited as Patrick, "Measurement of Water Quality."

35. Ibid.

36. Ruth Patrick, T. Bott and R. Larsen, *The Role of Trace Elements in Management of Nuisance Growths.* EPA 660/2-75-008, April 1975.

37. Harold H. Haskin, Rutgers University, to the author.

38. Ibid.

39. P.D. Uttomark and J.P. Wall, *Lake Classification—A Trophic Character-ization of Wisconsin Lakes*, EPA-660/3-75-033, June 1975.

40. Bostwick H. Ketchum, "Biological Implications of Global Marine Pollution," contribution no. 2974, Woods Hole, Woods Hole Oceanographic Institution. Hereafter cited as Bostwick, "Implications of Pollution."

41. Paul D. Uttomark, J.D. Chapin and K.M. Green, *Estimating Nutrient Loadings of Lakes from Non-Point Sources* (Madison, Wisconsin: University of

Wisconsin, 1974). EPA-660/3-74-020. Hereafter cited as Uttomark et al., *Estimating Nutrient Loadings.*

42. Ibid.

43. Ibid.

44. Uttomark et al., *Lake Classification.*

45. P.J. Dillon, *The Application of the Phosphorus Loading Concept to Eutrophication Research* (Burlington: Ontario: Canada Center for Inland Waters, undated).

46. Ibid.

47. G.M. Hornberger, M.G. Kelly and T.C. Lederman, *Evaluating a Mathematical Model for Predicting Lake Eutrophication* (Blacksburg, Va.: Virginia Water Resources Research Center, September 1975).

48. "Comprehensive Management of Phosphorus Water Pollution" EPA 66015-74-010, February 1974, p. 31.

49. Ketchum, "Implications."

Chapter 4
Urban Runoff and Nonpoint Sources

1. S.R. Weibel et al., "Urban Land Runoff as a Factor in Stream Pollution," *Journal of Water Pollution Control Federation* 36 (1964):914.

2. Whipple, "BOD Mass Balance."

3. E.H. Bryan, "Quality of Stormwater Drainage from Urban Land." Paper presented at Seventh Conference of the American Water Resources Association, 28 October 1971.

4. William T. Howard and R.D. Buller, "Unsewered Development and Water Quality," in *Urbanization.*

5. R.A. Young and K.L. Leathers, "Saline Irrigation Return Flows and the Economic Impacts of the Zero Pollution Discharge Objective." Eleventh American Water Resources Conference, Baton Rouge, November 1975.

6. U.S. Department of the Interior, *The Cost of Clean Water*, vol. 1, 1968.

7. Mark Pisano, "The Federal Perspective on Non-Point Source Water Pollution Control." Paper presented at the Water Pollution Control Annual Conference, Miami Beach, October 1975. Hereafter cited as Pisano, "Federal Perspective on Control."

8. Forrest C. Neil, "Urban Stream Management," in *Urban Runoff Quantity and Quality* (New York: American Society of Civil Engineers, 1975) pp. 14-16.

9. K.E. McIntyre, "The U.S. Army Corps of Engineers and Water Quality Control," in *Urbanization.*

10. James Wright, "Future Consideration for Assimilative Capacity Management of the Delaware River Estuary," in *Urbanization.*

11. D.A. Rickert, W.G. Hines and S.W. McKenzie, "Planning Implications of Dissolved Oxygen Depletion in the Willamette River, Oregon," in *Urbanization*. Hereafter cited as Rickert et al., "Planning Implications of Oxygen Depletion."

12. Pisano, "Federal Perspective on Control."

13. *National Water Quality Inventory: 1975 Report to Congress*, Environmental Protection Agency, EPA 440/9-75-014, 1976. Hereafter cited as *Water Inventory: Report.*

14. J.S. Zogorski et al., "Temporal Characteristics of Stormwater Runoff—An Overview" in *Urbanization*.

15. D.V. Dunlap, "The Influence of Precipitation on Biochemical Oxygen Demand on Nine Streams in New Jersey." Doctoral thesis, State University of Rutgers, January 1976, 121 pages.

16. William Whipple, Jr., J.V. Hunter and S.L. Yu, "Characteristics of Urban Runoff—New Jersey, *Water Resources Research Institute*, July 1976. Hereafter cited as Whipple et al., "Characteristics of Urban Runoff."

17. Ibid.

18. Enviro Control, Inc., *Total Urban Water Pollution Loads: The Impact of Stormwater*, (Report for the Council on Environmental Quality, 1974, PB. 231-730).

19. *Staff Draft Report: Water Quality Analysis and Environmental Impact Assessment of PL 92-500*. Technical Volume, National Commission on Water Quality, p. III-E-19 and III-E-9.

20. S.L. Yu, William Whipple, Jr., and S.L. Hunter, "Assessing Unrecorded Pollution from Agricultural Urban and Wooded Lands," *Water Research* 9 (1975): 849-52 (Pergamon Press).

21. N.V. Colston and A.N. Tafuri, in *Urbanization*.

22. N.V. Colston, "Characterization and Treatment of Urban Runoff," EPS-670/2-74-096 (1974), U.S. Environmental Protection Agency, Washington, D.C.

23. D.G. Shaheen, *Contributions of Urban Roadway Usage to Water Pollution*, EPA 600/2-75-004, March 1975.

24. Robert M. Ragan and A.J. Dietemann, "Impact of Urban Stormwater Runoff on Stream Quality," in *Urbanization*.

25. William Whipple, Jr., J.V. Hunter and S.L. Yu, "Unrecorded Pollution from Urban Runoff," *Journal of the Water Pollution Control Federation*, May 1974, pp. 873-885.

26. "The Costs of Sprawl." A report to the Council on Environmental Quality, Housing and Urban Development and the Environmental Protection Administration, Real Estate Research Corporation, April 1974.

27. W.G. Wilber and J.V. Hunter, "Contribution of Metals Resulting from Stormwater Runoff and Precipitation in Lodi, N.J." in *Urbanization.*

28. *Water Inventory: Report to Congress*, EPA.

29. DiGiano et al., "Projection of Pollution Effects."

30. Wilber et al., "Report to the Council."

31. Whipple, *Characteristics of Urban Runoff.*

32. Ibid.

33. *Water Inventory: Report to Congress*, EPA.

34. T.J. Tuffey and F.B. Trama, "Temporal Variations in Tributary Phosphorous Leads," in *Urbanization.*

35. Clifford W. Randall, et al., "Characterization of Urban Runoff in the Occuquan Watershed in Virginia," in *Urbanization.*

36. Uttomark et al., *Estimating Nutrient Loading.*

37. *Water Inventory: Report to Congress*, EPA.

38. Dr. Lowell Douglas to the author, at Rutgers University, New Brunswick, N.J.

39. Uttomark et al., *Estimating Nutrient Loading.*

40. *Water Inventory: Report to Congress*, EPA.

41. J.V. Hunter, S.L. Yu and W. Whipple, Jr., "Measurement of Urban Runoff Petroleum," in *Urbanization.* Hereafter cited as Hunter et al., "Runoff Petroleum."

42. Michael D. LaGrega and John D. Kennan, "Characterization of Water Quality from Combined Sewage Discharges," in *Urbanization.*

43. Herbert G. Poertner, *Practices in Detention of Urban Stormwater Runoff*, special report no. 43, American Public Works Association, 1975.

44. Donald L. Feuerstein and Alan O. Friedland, "Pollution in San Francisco's Urban Runoff," in *Urbanization.*

45. Alan O. Friedland, "Citywide Master Planning," in *Urban Runoff Quantity and Quality* (New York: American Society of Civil Engineers, 1975, pp. 8-13).

46. F.C. Neil, "Water Quality—The Necessity for Planning," in *Urbanization.*

47. William C. Pisano, "Cost Effective Approach for Combined and Storm Sewer Cleanup," in *Urbanization.*

48. John A. Lager, "Storm Water Treatment: Four Case Histories," *Civil Engineering,* December 1974, p. 40.

49. Richard Field and John A. Lager, "Urban Runoff Pollution Control—State-of-the-Art," *Environmental Engineering Journal*, vol. 101, EEI, February 1975. American Society of Civil Engineers.

50. McIntyre, "Corps of Engineers and Water Quality Control."

51. R.P. Shubinski, "Stormwater Treatment vs. Storage," in *Management of Urban Storm Runoff*, Urban Water Resources Research Program Technical Memo no. 24, American Society of Civil Engineers, May 1974.

52. Enviro Control, Inc., *Impact of Storm Water.*

53. William Whipple, Jr., "Water Quality Planning for the Delaware Estuary—An Example," *Water Resources Bulletin*, vol. 11, no. 2, April 1975. Hereafter cited as Whipple, "Planning for the Delaware."

54. William Whipple, Jr., J.V. Hunter, F.W. Dittman and G.E. Mattingle, *Oxygen Regeneration of Polluted Rivers: The Delaware River.* Environmental Protection Agency, December 1970, 16080 DUP.

55. William Whipple, Jr., "Mobile Oxygen Dispersion Craft," *Water Resources Bulletin*, American Water Resources Association, vol. 9, no. 4, August 1973, pp. 639-646.

Chapter 5
Municipal and Industrial Wastewater Treatment

1. Eric D. Bovet, *Evaluation of Quality Parameters in Water Resource Planning*, Report to U.S. Army Engineering Institute for Water Research, December 1974.

2. R.G. Eilers, *Condensed One-page Cost Estimates for Wastewater Treatment*, Environmental Protection Agency, November 1970.

3. D.B. Porcella et al., *Comprehensive Management of Phosporous Water Pollution*, Environmental Protection Agency 600/5-74-010.

4. Clifford S. Russell, *Residuals Management in Industry: A Case Study of Petroleum Refining*, (Baltimore: Johns Hopkins University Press, 1973).

5. Clifford S. Russell, in *Pollution Prices and Public Policy* by Allen V. Kneese and Charles L. Schultze, Brookings Institute, 1975.

6. A. Wolman, "Industrial Water Supply from Processed Sewage Treatment Plant Effluent at Baltimore, Md.," in *Water, Health and Society* (Bloomington: Indiana University Press, 1969, pp. 74-83).

7. News Release of the Dow Chemical Company, 27 August 1975. In *Chemical Engineering, Chemical Week* and other journals and as a communication from the company, April 1976.

8. Hunter et al., "Measurement of Petroleum.

9. News release, Interstate Commerce Commission, 18 July 1975.

Chapter 6
Principles of Water Quality Planning

1. *Multiple Objective Planning of Water Resources*, volume 1, Idaho Research Foundation, Inc., 1974.

2. "Report by Commission on Water Quality."

3. Bovet, *Evaluation of Parameters in Planning.*

4. Michael B. Sonnen, "Different Economics for Water Quality Planning," ASCE Specialty Conferences on Water Resources Planning and Management, Fort Collins, Colorado, July 1975.

5. David H. Howells, "An Assessment of Water Resources Research," Interstate Conference on the Environment. U.S. Water Resources Council, Cincinnati, August 1974.

6. Kneese and Bower, *Managing Water Quality.*

7. U.S. Water Resources Council "Principles and Standards for Planning Water and Related Land Resources, *Federal Register*, vol. 38, no. 174, pt. 3, 10 September 1973.

8. R.E. Pfister and R.E. Frenke, "The Concept of Carrying Capacity. Its Application for Management of Oregon's Scenic Waterway System." *Rogue River Study—Report 2*, (Corvallis: Oregon State University Press, June 1975) for the Oregon State Marine Board and Water Resources Institute.

9. M.L. Warner and D.W. Bromley, *Environmental Impact Analysis: A Review of Three Methodologies.* Water Resources Center (Madison, Wisconsin: University of Wisconsin, May 1974).

10. W. Schindler, "The Impact Statement Boondogle," *Science*, vol. 192 4239, 509, 7 May 1976.

11. Russell W. Peterson, "The Impact Statement—II," *Science* 193:4249, 16 July 1976.

12. Section 101 PL 89-298 Rivers and Harbors Act of 1965.

13. McIntyre, "Corps of Engineers and Water Control."

14. Merrimack Wastewater Management. Summary volume and 7 appendixes, New England Division, U.S. Army Corps of Engineers, November 1974.

15. Ibid.

16. Michael B. Sonnen, "Cost Implication of Current Water Quality Standards," paper prepared for an Urban Water Research Workshop, Quail Roost, N.C., July 1975.

17. U.S. Department of the Interior. Delaware Estuary Comprehensive Study. (Preliminary). Philadelphia: U.S. Department of the Interior, 1966.

18. Rickert et al., "Planning Implications."

19. *Formulation and Use of Practical Models for River Quality Assessment*, Geological Survey Circular 715-B.

20. *Draft Guidelines for Areawide Waste Treatment Management Planning.* Washington, D.C., EPA, October 1974.

21. Ibid.

22. Natural Resources Defense Council Inc. "Docket," May 1, 1975.

23. U.S. District Court for the District of Columbia Civil Action No. 74-1485, memorandum and order filed 5 June 1975.

24. Draft Guidelines, EPA.

Chapter 7
Water Quality Control in Other Countries

1. Kneese and Bower, *Managing Water Quality.*

2. Ibid.

3. Ibid.

4. William Whipple, Jr. et al., *Instream Aeration of Polluted Rivers*, Water Resources Research Institute, (New Brunswick, N.J., State University of Rutgers, 1969).

5. W. Whipple, Jr., "Oxygen Regeneration of Polluted Rivers: The Delaware River," Water Pollution Control Research Series, Environmental Protection Agency, Water Quality Office, 16080 DUP, 1970.

6. W. Whipple, Jr., "Oxygen Regeneration of Polluted Rivers: The Passaic River," Environmental Protection Agency, Water Quality Office, 1971.

7. Whipple, "Oxygen Craft."

8. Kneese and Bower, *Managing Water Quality.*

9. Toms, "Water Control in Great Britain."

10. Kneese and Bower, *Managing Water Quality.*

11. D.A. Okun, "Strategies for Assessing High Quality Drinking Water in the Future," in *Urbanization.*

12. Toms, "Water Control in Great Britain."

13. Ibid.

14. Ibid.

15. P.F. Teniere-Buchot and F. Fiessinger, "The Organization and Management of Water in France: Public and Private Aspects," in *Urbanization.*

16. Ibid.

17. Ibid.

18. Ibid.

19. *The Development of Water Basin Agency Action During Plan 4.* Minister of the Environment, France, 1972.

20. Ibid.

21. *Osaka and Its Public Works*, Bureau of Public Works, City of Osaka, Japan.

Chapter 8
Water Pollution Control Strategy and Combined Planning

1. Patrick, "Thoughts Concerning Measurement."

2. U.S. Water Resources Council, "Principles and Standards for Planning

Water and Related Land Resources," 38, *Federal Fegister*, no. 174, Part 3, 10 September 1973.

3. William Whipple, Jr., "Water Quality: A Re-Evaluation," in *Urbanization.*

4. Whipple, *Urban Runoff.*

5. Whipple, *Urbanization and Water Quality Control.*

6. Water Resources Management for Metropolitan Washington, Washington, D.C.: Metropolitan Washington Council of Governments, December 1973.

7. B.B. Berger and J.A. Kusler, "Lakeshore Zoning to Control Non-Point Sources of Pollution," in *Urbanization.*

8. Staff Draft Report, Commission on Water Quality.

9. Neil, Forrest C., "Urban Stream Management" in "Urban Runoff, Quantity and Quality," American Society of Civil Engineers, pp. 14-16, 1975.

10. Whipple, "Planning for the Delaware."

11. Enviro Control, Inc., "Impact of Storm Water."

12. Staff Draft Report, National Commission on Water Quality.

Appendix B
Technical Note

1. V. Pareto, *Manuel d'Economice Politique*, 2d ed., (Paris: Girar, 1927).

2. N. Kaldor, "A Note on Tariffs and Terms of Trade," *Economica* 7, November 1940.

3. T. Scitovsky, "A Note on Welfare Propositions in Economics" *Review of Economic Studies*, vol. 9, November 1941.

4. I.M.D. Little, *A Critique of Welfare Economics*, 2d ed., (New York: Oxford University Press, 1957).

5. W.J. Baumol, *Economic Theory and Operations Analysis* (Englewood Cliffs, N.J.: Prentice Hall, 1961).

6. Ibid.

Glossary

Advanced waste treatment: Treatment more thorough than the usual secondary waste treatment. The 1972 Act does not specify goals in these terms, but instead uses new definitions such as "best practicable" and "best available" treatment.

Algae: the microscopic plant life which, with bacteria, are the most abundant forms of life in water

Anaerobic: a condition where no appreciable dissolved oxygen exists. Also the designation given to bacteria which live in such conditions. Used in contrast to aerobic.

Aquifer: underground porous layers used as a source of water supply, usually by means of wells

Artificial aeration (or artificial oxygenation): the process of adding additional oxygen to a waterway by bubbling air (or oxygen) through it, or mechanically agitating it in such a way that it absorbs oxygen more rapidly from the atmosphere

Automatic samplers: devices which are left unattended, and which periodically withdraw samples of water for retention until collected

Benefit-cost approach: This was the method specified by Congress during the 1930s, by which proposed projects had both their costs and their benefits evaluated, both reduced to an annual basis. Only those projects were to be authorized for which the evaluated benefits exceeded the costs.

Benthal fauna: bottom creatures, including Mayfly larvae, snails, etc.

Benthal oxygen demand: oxygen demand exerted from the bottom of a stream, usually by biochemical oxygenation of organic material in the sediments

Bioassay: use of organisms to determine the biological effect of some substance, factor, or condition; testing of toxicity by scientifically conducted experiments on living creatures

Biochemical oxygen demand: The potential demand for oxygen which will be needed for oxidation of organic materials in the water by a combination of bacterial and chemical action. Unless qualified, the demand to be exerted within 5 days is meant. This is the most commonly used indicator of pollution.

Biological community: An interrelated group of different species occupying a given

BOD: biochemical oxygen demand

Brackish water: waters containing only part of the salt content existing in seawater

Cfs: cubic feet per second

COD: chemical oxygen demand

Coliform counts: numbers of bacteria of the types found in human digestive systems

Combined sewers: Sewers which convey both storm water runoff and the overflow during storms from sanitary sewers. A major source of unrecorded pollution in most cities.

Communities: biologically a community is an interrelated group of different species occupying a given space

Combined planning: Planning which includes a number of functions rather than only one. That is, it provides for solving problems related to water quality, water supply and sometimes flood control in combination rather than considering them independently of one another. Combined planning may be either integrated or coordinated.

Comprehensive planning: planning which supposedly considers all of the possibilities of water resource development rather than only a specific objective of direct concern, such as flood control

Conservationists: those environmentalists who emphasize wise and restrained use of our national resources, the more traditional environmental view

Coordinated planning: Planning conducted by different approaches to the various subjects considered, but related to one another so as to obtain the advantages of a combined plan, as far as practicable. Coordinated planning implies a cooperative approach between two or more planning units.

Corps of Engineers: Part of the U.S. Army. Besides military functions, is charged with national programs of flood control, navigation and related water supply. Known as the nation's largest builder of dams. Carries on comprehensive water quality control studies when so directed by Congress. Controls all filling and all channel alteration on U.S. rivers.

Design storm: storm selected for planning or design purposes as typical of weather contingencies to be guarded against

Dissolved oxygen: The oxygen dissolved in water. It is an important indicator of water quality. Measured in milligrams per liter (mg/1), which may be expressed parts per million (ppm).

DO: See dissolved oxygen.

Economies of scale: the lower unit costs of mass production or other large scale operations

Ecological impact: the total effect of an environmental change, either natural or manmade, on the ecology of an area

Ecosystems: Interacting complexes of living organisms and their physical environment. May include one or more biological communities.

Effluent: the wastewater discharged by an industry or municipality after treatment

Effluent charges: a system under which control of pollution is exercised by making a charge for the amount of pollution entering the stream, the charge being related to the damage done by that quantity of pollutant further downstream

Environmental Protection Agency: The federal agency charged with implementation of Federal water pollution control legislation (with minor exceptions). Also charged with Federal air pollution control and solid waste programs.

Estuary: Area where freshwater meets saltwater, e.g., bays, mouths of rivers, salt marshes, and lagoons. Estuaries serve as nurseries and spawning and feeding grounds for large groups of marine life and provide shelter and food for birds and wildlife.

Eutrophication: The natural process of aging of lakes, by which organic matter is formed by growth of algae and other plants. The process is accelerated by excess nutrients, resulting in unwanted algal blooms (heavy growths).

Fate: the changes in a substance once it enters a stream

Federal Water Quality Administration: predecessor of the present Federal Environmental Protection Agency

First flush effect: in a storm sewer the (usual) condition when pollutants are heavily concentrated in initial runoff at the onset of a storm

Flow: proportional composite sampling: A sample is taken at each time interval, but portions of each sample are then combined on the basis of flows; that is, the part used of a sample at flow x is x units, and the part used of the sample at flow y is y units.

Heavy metals: Important constituents in stream pollution. Most usual metals to cause difficulties are lead, zinc, copper, nickel, chromium and mercury. They are often toxic, and persistent.

Integrated planning: A single planning approach applied uniformly to the various subjects considered, to obtain the best plan. In practice an integrated plan is prepared by a single organization.

Intangible effects: effects (of a project) which could not be evaluated in dollar terms

Limiting nutrient: there is usually one nutrient in limited supply for algal growth, the others being present in sufficient quantity

Loading of pollutant: The mass passing a given point, in pounds per unit of time. Obtained by multiplying the pollutant concentration, times the discharge, times a constant. Loading varies directly as the product of concentration and stream flow.

Low flow augmentation: the process of supplementing low flows of a river, usually by releases of reservoir storage

Man: often used in this text in the dictionary sense of mankind in general, including both male and female, with no suggestion implied as to relative superiority of one or the other, in any context

Mechanical aeration: a form of artificial aeration

Mg/l: milligrams per liter, equivalent to parts per million by weight of the substance in water

Modeling: The mathematical simulation of natural processes in waterways; although usually physical and chemical processes, biological activities may also be simulated. The model is applied to analysis of hypothetical future conditions.

Multiple objective planning: planning which considers more than one general national objective, i.e., both economic efficiency and environmental quality

Multiple-purpose project: one designed to achieve more than one purpose such as both flood control and water supply

NCWQ staff: Staff of the National Commission on Water Quality, in reports published in the fall of 1975.

New York Bight: The wedge shaped extension of the Atlantic Ocean lying between the beach coast lines of New Jersey and Long Island. The point of the bight leads to New York City harbor.

Nitrification: the process by which ammonia is changed first to nitrites and then to nitrates, by bacterial action, consuming oxygen in the process

Non-structural alternatives: programs (in connection with water pollution control) which consists of approaches other than the structural approaches of constructing waste treatment plants, etc.

NPDES permits: FPA permits specifying conditions of compliance with federal law as to allowable wastes to be discharged

Nutrients: Substances necessary for growth of algae and bacteria in water, such as nitrates and phosphates. There are many others.

Optimal: the best available alternatives for the purpose or purposes intended

Organic: Referring to or derived from living organisms. In chemistry, any compound containing carbon. Domestic and many industrial wastes are largely organic.

Oxygen sag curve: the characteristic curve representing the variation of dissolved oxygen concentration in a stream below a major source of pollution

Photosynthesis: the basic process of plant life, by which carbon dioxide and water, in the presence of sunlight and nutrients are converted to organic substances, with liberation of oxygen

Pollution concentration: the proportionate amount of pollutant in water, usually measured in milligrans per liter

Pollution: see water pollution

Pollution loading: the total weight of pollutant carried by a stream per unit of time

PPM: parts per million

Preservationists: those environmentalists who would preserve nature in its "original" state, substantially unchanged by man

Rating curve: on streams this means the graphical expression of the relationship of height of the water surface on a gage to the discharge of the stream

Receiving waters: streams or other waterways which receive polluting discharges

Recharge: percolation of water into an aquifer from the ground surface bodies of water

Residence time: the length of time a given mass of water remains in a lake (or water body)

Residuals management approach: the analysis of water pollution, air pollution, and solid wastes problems on a single basis, recognizing that wastes may be transformed from one form to another, but in the end must be disposed of

Salt water intrusion: salt water, being heavier than fresh, tends to flow under it and displace it, by penetrating porous formations from the coasts, especially those deep down

Secondary waste treatment: The form of treatment usually given to sewage and organic industrial wastes. The process has physical and biological stages. It includes the processes referred to as primary treatment.

Section 208 planning: areawide planning of water pollution control under provisions of the 1972 Water Pollution Control Act

Sediment: popularly referred to simply as "silt," but clays, sands and organic substances are important components

Sludge: The solid matter removed from waste water by means of treatment. Sludge handling involves the processes that remove solids and make them ready for disposal.

Shadow prices: estimates of money value applied to items for which no market value exists

TBA: 2.3-6 trichlorobenzoic acid

Thermal pollution: degradation of water quality by the introduction of a heated effluent, which is usually the result of the discharge of cooling waters

TOC: total organic carbon

Toxicity: the characteristic of being poisonous or harmful to plant or animal life

Turbidity: cloudiness, lack of clarity in water

Urban runoff: term usually used to designate all pollution emanating from an urban area except known point sources of untreated wastes and treatment plant effluents

Waste load allocation: an administrative ruling as to the maximum amount of polluting substances allowed to enter a stream from a given facility

Water pollution: the presence in water of chemical, physical, biological, and radiological characteristics detrimental to its intended use

Watershed: land draining into a stream used lower down for water supply

Wetlands: Swamps and marshes. They usually have unsuspected ecological value.

Index

Index

About the Author

William Whipple, Jr., has been director of the Water Resources Research Institute at Rutgers University for eleven years, working mainly on water pollution, environmental impacts, and the planning of remedial programs. A graduate with honors from West Point, he was sent to Oxford as a Rhodes Scholar, where he studied economics, politics and philosophy, and he finished his education by obtaining a graduate degree of civil engineering from Princeton University.

In his earlier life he served in the Army Corps of Engineers, including planning and construction of flood control, navigation, water supply and hydroelectric power projects, and retired with the rank of brigadier general. His past experience includes service as President of the American Water Resources Association, Chairman of the Urban Water Resources Research Council, A.S.C.E., and Chief Engineer, New York World's Fair 1964-65. General Whipple has been officially commended by the President of the United States, and he is currently chairman of the Universities Council on Water Resources.